Must Be Relevant to Make Some Money
by
Dr. Phil Copeland

Must Be Relevant to Make Some Money
Copyright © 2016 by Dr. Phillip E. Copeland

ISBN (978-0-9982704-0-1)

CONTENTS

INTRODUCTION

You must be relevant to make some money. It doesn't matter if you work for a business or you're a business owner. Employee success depends upon relevance in the eyes of their bosses. The worker who brings genuine value to their employer will strengthen the overall capabilities of their organization. As their tacit knowledge continues to propagate around the DNA of the company, the employee will indeed grow more relevant and often flourish to higher levels. Employer success depends upon relevance in the eyes of their customers. Companies must offer a product or service of value to consumers that appears relevant enough to entice their interests and convert these consumers into loyal customers.

No magic formula will guarantee you relevance, leading to your success. However, some essential ingredients can spice up your odds. Key fundamentals include how you notice trends, achieve a competitive advantage, innovate things or ideas, leverage the right business model, make ethical decisions, act socially responsible, effectively manage risks, adopt mobile technologies, go global, and lead interdependently. I will elaborate on these quintessential principles throughout the chapters of this book. You will notice that they overlap each other quite frequently.

The content in this publication should enable you to focus less on problem solving issues that will close a deficit gap and more on achieving extraordinary results that will close the abundance gap. This is how successful people and organizations accomplish impossible goals, so as they seem. You will not only appreciate the material showcased in this book, but your behavior will be swayed by such knowledge. Much of this content may already reside dormant inside you and you will say to yourself throughout this publication journey, "Ahha!" You increasingly feel the desire to transform such newfound applicable familiarity into actionable results, to become relevant and consequently make some money.

CHAPTER ONE:
MEGA TRENDS

You must be relevant to make some money. A trend is the general direction in which something is developing or changing, such as in the fashion industry with the hottest styles. Trends tend to dictate relevance. Some of these trends include innovation and mobile technologies. However, by the time a trend is noticed, there could already have been a missed opportunity to gain a competitive advantage in the market. Trends start out with a novelty (something new and original) that is less noticeable and harder to detect. Novelties more often become fads and less often become trends (Schwarz, Kroehl, & von der Gracht, 2014).

Managers who are capable of detecting the weaker signals of a novelty will jump ahead of the rest before it becomes more obvious as a trend. Trends have a much longer lifespan than fads and possess the potential to influence business markets in the long term. Upon analysis, a determination of patterns can be used not only to understand past events and present situations, but to anticipate possible future business conditions. Various trends, often unrelated at the moment, can actually converge into more of an overarching global direction known as a mega trend.

Mega trends are the underlying transformative currents driving trends and influencing the global expanse of activities and processes shaping society. The key is for businesses to be aware of mega trends and to recognize any potential convergences of unrelated trends that could make what currently appear to be irrelevant trends for businesses to become relevant for them in the future. It would be nice to have a crystal ball or special talents to predict the future accurately, but since such tools are unlikely weapons, the more viable option is to recognize how the patterns of mega trends could change the overall business landscape in the future.

Some astute observers in today's business world have identified a handful of mega trends as significant determinants to reshape the landscape in the next 15 to 20 years. Such trends include individual empowerment, diffusion of power, sustainability, and technological evolution. The most important of these trends is individual empowerment, which is infused throughout most of the other trends. Within the next few decades, the spirit of individual initiative will help to diminish poverty, build a global middle class, expand access to education and improve healthcare. The diffusion of power is a trend showing that by the year 2030, Asia will have surpassed both North America and Europe as the new top global power in regards to their gross domestic product and other factors such as their immense population, technological investment, and military expenditures (Marien, 2013).

Past trends have depended upon their long-term sustainability, but now sustainability in itself is considered a mega trend. It will become necessary to identify influential stakeholders who will influence an organization. Managers will need to determine how the organization will affect these influential stakeholders. Expectations of ominous conditions will impel leaders to discover original ideas and leverage them in creative and innovative ways to find opportunities and remedy any issues such as escalating cost of operations or declining revenues. They will need to engage with internal and external stakeholders in forming strategic alliances with other companies (Tideman, Arts & Zandee, 2013).

The technological mega trend shows promise to foster a healthier society. Dr. Diego Miralles, head of Janssen Healthcare Innovation, emphasized that empowering and trusting consumers with their own information could enable them to thwart many of the preventable chronic diseases. The technologies of the internet and mobile channels enable consumers to acquire more health-related information, leading to better decisions about their health and the intervention of any health issues. Laborers in developing countries may eventually foster a healthier physical well-being through this technological trend. The expansion of a healthier workforce and the increased longevity of those workers will contribute to the overall sustainability threaded throughout other trends.

When managers make decisions, they should consider the current conditions affecting their businesses and their industries. But for the sake of sustainability, they should also examine potential future conditions. Managers can forecast the future possibilities if they know the number of quantitative input and output variables, but these variables are limited to a single outlook of the future. During an extreme transformative period, which could present more chaotic, uncertain direction, managers may need to discard forecasting techniques and embrace a multiple scenario analysis. Forecasting focuses on the outcome, while scenario analysis examines the dynamics and processes involved.

Scenario Planning

A multiple scenario analysis examines realistic versions of how a competitive landscape could evolve in the future. The fluid interaction of events, conditions, and changes will mold potential scenarios for managers to scrutinize, paying close attention to cause and effect. Once a set of input and output variables has been established within a given scenario, then forecasting techniques can then be applied.

The process of entertaining a variety of future realistic business conditions can help reveal hidden shadow beliefs that may have biased past management perspectives and decisions. Managers can leverage the new knowledge acquired from scenario planning to develop strategic plans that will better prepare the organization for what could happen and how leaders within the organization will react. Strategies will need to be flexible enough to adjust for any of the probabilistic scenarios.

The primary purpose of scenarios should be to identify the determinants for each set of future business conditions and to understand how they work. The actual accuracy of the scenarios in regards to the prediction of events and trend assessments would be beneficial for the organization, but they are only of secondary importance. A key component for the development of scenarios is to evaluate how factors such as technology, government, and workforce demographics may influence key business factors and subsequent operational needs. Brainstorming all possible "what if" situations will help managers to expand their imagination of future conditions. Managers could perform a SWOT analysis to evaluate the strengths, weaknesses, opportunities, and threats for each scenario that has been presented. A plan should be established for each scenario that compares the current conditions with the new ones and to identify any gaps, along with ways to close them.

Scenario and Mega Trend Model

A scenario and mega-trend model (See Figure 1) presents a standard process that a business can use to address the mega trends affecting the organization and its industry. The trend analysis will reveal patterns that can be helpful in creating scenarios with potential inputs, outputs, and outcomes in a given competitive market situation.

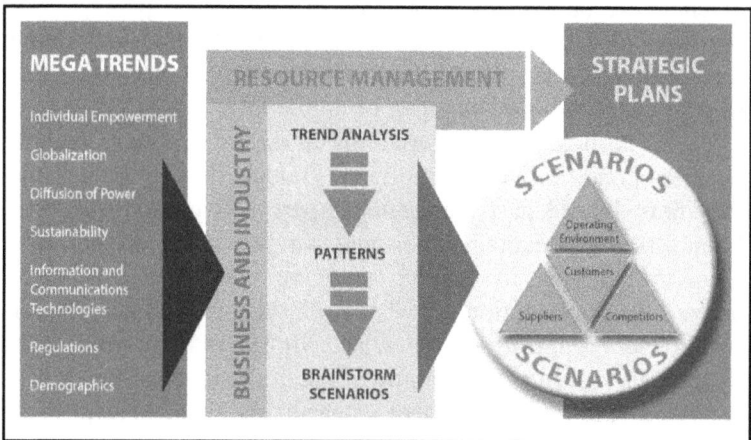

Figure 1. Scenario and Mega-Trend Model. Elements to consider in strategic planning for any given set of potential business conditions to better prepare businesses for the future.

Inputs include competitive drivers such as the overall market size, market growth rate, and market attractiveness factor. Outputs involve the responses from stakeholders such as consumers and distribution channels. Outcomes represent the responses from individual competitors within the context of other factors. Scenarios involve any given combination of operating conditions, suppliers, customers, and competitors.

This model should create a clearer perspective for leaders to better anticipate potential future conditions and determine the best course of actions, resulting in more informed decisions if ever faced with one of the possible scenarios. Resource management will have to be kept in mind throughout the entire process of developing strategy both with and without consideration to any scenarios. The challenge will be to balance resources and efforts. The goal is to ensure that businesses avoid missing any future opportunities and being caught by surprise from any potential future risks. At the same time, business leaders must not overspend time and resources on possible conditions that may not be realistic and probable.

Case Analysis: Exxon Mobil

Exxon Mobil is currently the top leader within the oil and gas operations industry and ranked sixth overall on the Forbes world's biggest public companies. Mega trends indicate a major shift will occur in the current competitive landscape during the next few decades. The mega trends that are considered in the proposed scenario and mega-trend model include individual empowerment, globalization, sustainability, diffusion of power, information and communications technologies, regulations and demographics.

Trend analysis indicates that patterns suggest the following future conditions. Individual empowerment could reduce poverty and lend to the growth of a global middle class. The global population will have increased from 7.1 billion people in the year 2012 to 8.3 billion in the year 2030. Globalization depends on long-term sustainability, which is now considered a mega trend. The identification of influential stakeholders and the use of partnerships will become necessary tools for a leader. They will need to find creative, innovative and original approaches to remedy problems and find opportunities. The diffusion of power is another significant mega trend that emphasizes that before the year 2030, Asia will surpass the combination of both North America and Europe in regards to global power derived from gross domestic product, population size, military spending and technological investment.

The evolution of information and communications technologies will foster the global adoption of the internet and mobile devices, which will promote a healthier behavior around the world, to include developing areas. The growing use of the hydraulic fracturing of rock is stimulating the flow of wells for natural gas resources and will significantly expand the use of natural gas reserves and increase crude oil productions, which could cause the collapse of crude oil prices. Demographic trends for the next few decades include: mature countries will struggle to maintain living standards, there will be a decline of youthful societies and the level of urbanization in developing countries could equal to the entire world history of total construction.

Exxon Mobile could face the following scenario by the year 2030: The shift of global power to the Asian sector will influence the operating environment now facing the collapse of crude oil prices, but the extension of natural gas resources. Political influence from a global level, change in workforce demographics and less resource investment will impinge upon operations. It will become necessary to decrease operating costs and focus on new revenue-generating areas other than crude oil productions.

The new global middle class will change the customer demographics. The competitive advantage will shift to Asian companies who will now dominate the oil and gas operations industry. The collapse of crude oil prices will change the scheme of suppliers from crude oil to more of natural gas resources and other alternative energy resources.

Exxon Mobil will need to identify the stakeholders who now benefit them the most and determine how to provide satisfactory value to them. Partnerships with organizations rich in resources may be beneficial if the course of action will be to remain with the current organizational mission and the added resources will rejuvenate capabilities.

A scenario response plan will need to be integrated into Exxon Mobil's strategic plan, while taking resource management into consideration. Exxon will need to balance the use of their resources in building capabilities that will prepare for possible, but realistic conditions.

CHAPTER TWO:
COMPETITIVE ADVANTAGE

You must be relevant to make some money. Company success depends upon relevance in the eyes of their customers. Organizations such as Exxon Mobil compete for their share of the consumer market and strive to gain any possible advantage over their competitors in accomplishing this objective. Some industries may maintain a single leader, while in other industries more than one company may have found a competitive edge, enabling them to surface towards the top. Different companies within the same industry can actually achieve and sustain a competitive advantage without directly disrupting one another. The ideology of market segmentation will explain how this phenomenon could occur. Competitive advantage will be discussed in further detail. We will examine examples of two leaders within the same industry, followed by a glimpse at the potential outlook of this industry and any recommendations to help the businesses continue into the future.

According to market segmentation theory, an industry reflects in essence an overall heterogeneous market environment with various distinct consumer preferences and immobile resources. This would also implicate dynamic pricing which reveals a difference in prices according to the consumer demographics and purchasing history. But, an industry is comprised of market segments which are groups with homogenous consumer preferences and mobile resources. These groups of commonality can be based on consumer demographics, geographic location, lifestyles or purchase behavior.

Businesses will be more realistically able to target a particular market segment with its distinct skills or resources that can potentially provide them a comparative advantage, leading to a competitive advantage within a particular marketplace and ultimately superior financial performance. More than one company within the same industry can potentially achieve an advantage over their competitors, but within different market segments.

Dr. Jay Barney defined a competitive advantage to be a value-creating strategy executed by an organization not yet employed by their competitors. A company can achieve a competitive advantage with skills or resources it asserts to offset its competition. Trained personnel with unique skillsets that produce a specialized product could catapult the company past its competition. Resources such as ownership of a distribution facility or bargaining power with outside distributors would lower costs within the distribution chain, enabling a company to gain the ability to offer a more economical product to consumers (Musetescu, 2013). Organizations should choose resources with comparative advantages to help achieve a competitive advantage.

Companies will need to exhibit perceived moral standards in efforts to foster a social trust in building a long-term relationship with consumers. Trust building will nurture a consumer commitment that will lend itself to a sustainable competitive advantage, positioning the organization into an exclusive strategic position within the marketplace. The sustainment of this competitive advantage will rely on distinct, immobile resources that cannot easily be obtained or copied by the competition.

A sustainable competitive advantage that cannot easily be copied or substituted by competitors will require a distinct combination of ingredients that may involve low cost strategies, product patents, recognizable brand name, great customer relationships providing excellent services, trained personnel with unique skillsets and investment in research and innovation. Competitors that can eliminate the cost of research and innovation by imitating the incumbent leader's product or process will be in a better position to offer a comparable product or service at a lower price and threaten their sustainable competitive advantage.

Incumbents will need to monitor their organizations continually, as well as the competition and the environment in general, along with any public policies and regulations for any potential issues that could potentially diminish the competitive advantage. Resource management decisions will apply any needed adjustments in resource allocations to defray potential threats or to build upon a competitive lead. Incumbents should assume competitors will respond with reactive innovation in hopes of surpassing their competitive advantage with either a product of more value, a cheaper price or both.

Successful organizations who sustain a competitive advantage have been able to foster an organizational culture, which inspires workers to innovate. However, they should be cognizant not to overstep innovative efforts into unfamiliar territory where they may lack available skills and capabilities to integrate the new ideas successfully (Dustin, Bharat & Jitendra, 2014). Incumbent industry leaders who have sustained a competitive advantage tend to accumulate assets and market power, providing an economies of scale to help keep them stay afloat under adverse conditions. The sustainable competitive advantage creates a superior value, allowing incumbents to demand higher premium prices, expand their market share and incur more value, ultimately realizing higher profits. However, the incumbent is usually not very adaptable to dynamic environments. Newer competition could overtake the incumbent through the means of disruptive innovation (de Brito & Brito, 2014).

Case Analysis: The Tale of Two Smartphone Giants

Apple was the first entrant into the smartphone market that set the pace for others to either keep up or pass. Upon their entry in the year 2007, Apple initially agreed to be carried for the first five years exclusively only by AT&T in return for percentage of profits that AT&T acquired from iPhone users. Apple preserved their control of the design, manufacture and marketing for the iPhone from the beginning to the present. The competitive advantage of the iPhone relies on their brand recognition, ease of use and their app market. An effective marketing strategy for the iPhone has branded its identity into the consumer minds since its inception. They have established a relationship with the consumer built on trust and loyalty to help sustain its competitive advantage (San-Martin & López-Catalán, 2013). The simplistic, sleek design still offers the easiest usability among all of the smartphones (Spaid & Flint, 2014). Apple apps for the iPhone dominate all other competitors. A current iPhone user may not want to switch to another smartphone company if they risk losing a desired unique app only offered by Apple (Mojica, Adams, Nagappan, Dienst, Berger & Hassan, 2014). Apple mainly targets the market segment consisting of homogenous consumers who prefer the attractive and easy to use design of the Apple products.

Google stumbled into the smartphone market with a cumbersome Android model. Since the inception of the Android in the year 2007, continuous incremental improvements have added new features and resolved bugs from the older models to help improve usability for consumers. Newer versions of the Android have accelerated their slower beginning, surprising their Apple rival. The advancements of the Android have now surpassed the iPhone as the top leader in the smartphone industry. The Android mobile operating system is a low-cost customizable system that has become popular among technology companies. An influx of developers and enthusiasts has adopted its open-source code, expanding the Android OS across a wide array of community-based projects. There are well over a million available apps for the Android. The Android will efficiently manage the stored apps and automatically suspend inactive apps to help preserve the limited battery life of the mobile device (Hamdaoui, Alshammari & Guizani, 2013). Google mainly targets the market segment consisting of homogenous consumers who prefer the advanced features offered on mobile devices.

The technology acceptance model (TAM) suggests that consumers will adopt a mobile phone that is easy to use and perceived to be useful for their needs. Sustainable competitive advantage will rely upon the adaptation to accommodate future conditions, while considering device improvements for increased ease of use and value-added features. As the level of curiosity with innovativeness among consumers continues to increase, so will the growth of mobile adoption and new or improved features of smartphones.

Dynamic Capabilities

Among today's competitive surroundings, the business organization accumulates the resources needed to provide them the capabilities to produce a commodity of value for the consumer. The organization will leverage this resource-based strategy to achieve a competitive advantage. Market conditions may warrant if the organization will need to change how they do business in order to sustain such an advantage. Their capacity to change the allocation of resources can dictate whether a current competitive advantage will be sustained for longer periods. This capacity is referred to as the organization's dynamic capabilities (Teece, Pisano, & Shuen, 1997).

Dynamic capabilities can be seen as the organization's ability to make prompt decisions to solve problems, seize opportunities and adjust their resources as needed in order to sustain a competitive advantage. The types of industries and environmental conditions may determine where these capabilities could prove beneficial. The decisive actions of leaders within more high-tech types of organizations will differ from those less attached to technology and innovation. Leadership direction for an organization competing within the global arena may require more flexibility than in local markets.

The spirited intensity within highly innovative-driven industries around the world has given rise for many companies to be flexible enough to change with fluid market demands. Despite the industry and environmental conditions, the allocation of resources is always critical in achieving a competitive advantage. An organization should be astute to their industry's profit potential in determining their strategy for the future, impacting how they will decide to allocate their resources. Porter's five forces model points out the determinants for the potential profit: entry barriers, threat of substitution, buyer's bargaining power, supplier's bargaining power and rivalry among industry incumbents. The organization's analysis can reveal how much competitive power they hold within the industry and whether they need to adjust their current use of resources to prepare for future conditions.

The organization will need to align their unique capabilities with consumer demands if they are to realize their greatest profit returns known as the "coherence premium." They will need to understand their value chain and establish flexible processes that keep focused on their primary objectives. The unique capabilities are comprised of the organization's core competencies in producing value for the consumer. Value may be a better product, better delivery of a product, faster execution of a product, a cheaper product or more reliability (Gupta, 2013).

The innovative management of a commodity, the quality of an organization's personnel and the internal controls used by management can all lend to the sustainment of a competitive advantage for an organization. Leaders will need to be aware of how change may be necessary and will bring about the need for an organization to learn new processes and rules. The culture within an organization will need to transcend their mindset from traditional routines to new practices (Gao, Tian, & Yu, 2014).

An organization's dynamic capabilities will capture opportunities and mitigate strategic risks. They are able to transform their capabilities when needed to meet new opportunities or to fix problems that may arise over the years. An evolutionary process will adjust their human, structural and intellectual capital in sustaining a competitive advantage (Shuen, Feiler, & Teece, 2014).

Leaders will need procedures in place to consistently monitor their organization's capabilities and scan the changing competitive landscape. They will benefit from an ability to foresee and secure new business opportunities. The greatest challenge may be how an organizational culture will adapt to change. Such change could force people to change jobs, learn new tasks or adjust to new roles. Professional development and specialized training will lend to them learning new skillsets (Wu, He, & Duan, 2013).

High-Tech Industry

The framework of dynamic capabilities initially contributed knowledge about the strategic agility needed by high-tech companies competing in high-velocity markets. These organizations often compete within an increasingly competitive environment. They face the challenge of recruiting skilled engineers from a limited talent pool. Organizations search for ways to change the industry through new technologies, energy resources, and cultivating a flexible talent pool.

Information technology (IT) may be a driving force of innovation, which can be the catalyst of sustaining a competitive advantage. An organization must fully understand any new technology that it is considering for integration into its infrastructure. Their knowledge of how the old and new elements are to interact will lend to a successful integration of how best to fit all of these elements together into a new system.

A strategy interlaced with dynamic capabilities will enable high-tech organization to better address rapidly changing conditions. Their organizational structure allows them to reconfigure their competencies and processes into unique capabilities of achieving and sustaining competitive and performance advantages. They will have the dual capacity of operating in both the mature and the emerging market segments.

Global Competition

The top companies that compete at a global level can respond expeditiously to a continually changing competitive landscape. A wealth of valuable assets can only be useful for an organization if they can effectively employ their resources to keep pace with any changes in consumer demand or to create a new market. The successful multi-national corporations know how best to manage their resources in sustaining a competitive advantage for both current and future conditions.

International diversification has extended the knowledge base and longitudinal perspective of the world's largest multi-national corporations. They are able to apply their dynamic capabilities to prolong their competitive advantage. These corporate giants remain at the top through their ability to expeditiously deploy any changes to continually provide value that the consumer needs. They maintain their position through technology, reputation or financial means.

These successful organizations will follow their own distinct path to obtain their position. Their current position will determine how much they decide for the present competitive environment and to what extent they can decide for future scenarios. Multi-national corporations that are more internationally diversified tend to expand their knowledge significantly, while magnifying their research and development efforts. They excel in innovation from the simultaneous investments in various countries and products.

The larger multi-national corporations have to concern themselves more about the overall complexity of the organizational structure versus the smaller organizations. Leaders should identify team compositions and establish a cross-sectional dynamics. Team members with diverse backgrounds can help expand their thought process to render more ideas. They should be empowered to pursue potential opportunities. An open positive atmosphere will enhance their collective creativity in discovering new innovative solutions or seizing lucrative opportunities (Fourné, Jansen, & Mom, 2014).

CHAPTER THREE:
INNOVATION THEORIES

You must be relevant to make some money. Innovation has been a key ingredient for organizations to sustain relevance for longer periods of time and seems to be one of the prominent buzzwords now in society. Everybody seems to act like this is some new hot thing, but we all know that innovation has been around since the beginning of time, just not so commonly highlighted as in the current mainstream versus the past.

The majority of businesses still tend to focus naturally on keeping up with their competition. They don't really explore innovative ways or ideas to invent something new for the consumer. This may be due to their limited resources or a decision driven by the organization's culture. Managers may be more comfortable with conventional strategies to build upon the existing market, instead of leaders looking for a new spice that would most certainly entice the hungry to want more of this delightful fresh flavor.

In the business world, innovation is a process intended to achieve a competitive advantage over competitors. Innovation can improve upon an existing product, service or the organization itself, with an incremental approach or create something new with a more radical approach. The different types of innovation are accompanied by different theories explaining the processes and implications of the innovation.

Disruptive Innovation Theory

Michael Raynor (2011) discussed how new ventures across all types of industries tend to succeed at the initial startup, if the new competitors focus on new markets or markets of little interest to the current successful incumbent. Once established within this market, if the new competitor is able to offer superior performance while using business models focused on the new or undesirable market that the incumbent can't replicate the same approach, then the incumbent faces a potential threat of losing major customers. This threat would now be considered a disruptive innovation.

However, if the new competitor attempts to use the incumbent's business model and solutions, also known as sustaining strategy, they increase the risk for failure. Clayton Christensen (1997) referred to this overall idea as a disruptive strategy. Disruptive innovation theory can be used to shape existing innovation ideas that can better deal with disruptive innovation.

The incumbent's ability to change and replicate the same approach as the new competitor will rely on factors such as their own resources, processes and values. These factors include resources, processes, and values of an organization. Even if the incumbent has the resources to change, it may need to adjust their current processes and values in order to be successful with change. The latter two factors could be of distinct advantages for the new competitor who poses the threat of disruptive innovation, if the incumbent cannot easily change the way they accomplish their tasks and determine their priorities.

For example, Digital Equipment Corporation did not survive the disruption innovation of the personal computer like their competitor, IBM. This was due to their inability to adjust their processes and priorities that IBM was able to adjust. Soren Kaplan introduced the use of a formal unique management tool referred as LEAPS to more effectively lead and manage disruptive innovation. This tool describes the basis for understanding how leaders create clarity during periods of uncertainty; leaders should appreciate that uncertainty presents as much opportunity as it does risks and they should push beyond the boundaries and comfort zone of their organization. Leaders should listen to themselves, instead of the market.

However, it is an assumption that the leadership of Digital Equipment Corporation more than likely did not use this approach. They appeared to have remained with their current successful processes and listened to the current market demands, instead of themselves, resulting in their inability to effectively cope with the disruptive innovation of the personal computer in the new market.

Maxwell Wessel and Clayton Christensen (2012) recommended that an organization, whether new or an incumbent, creates a disruption to increase the survivability of new markets. Disruption is described as more of a process versus an event that may happen over time, but may vary in the length of time. An organization should assess the disrupter's business model and their own strengths, and determine what elements would both help and hinder another organization, ultimately decreasing their current advantages. This would involve an organization to evaluate the extendable core of a competing organization and assess their actual capability to present a disruptive innovation.

Michael Rayner supported the idea that technological or business model advantages enable the extendable core. In reference to the personal computer disruption mentioned by Clayton Christensen and Michael Overdorf, Digital Equipment Corporation could not easily change their process from customizing to standardizing their computers. Their process tied into the actual customization for each customer and made it very costly to change to a new process that focused on standardization for overall customers. This presented a huge cost advantage for the disrupter, while creating a disruptive innovation that eventually led to the demise of Digital Equipment Corporation.

Successful organizations interpret opportunities as jobs they can accomplish for current or future customer while offering convenience and accessibility to the masses. Five types of barriers to counteract disruptive innovation (from easiest to the hardest) are mentioned: momentum, tech implementation, ecosystem, new technologies, and business model. A disrupter such as an online grocery store may be able to bring disruptive innovation upon a traditional grocery story with bulk shopping of staple items to realize cost savings. The online store will probably will only encounter the momentum barrier.

However, if their focus is on emergency goods that are needed now, this may force them to encounter the business model barrier with the serious implications of costly infrastructure changes that could involve the adaptation of the traditional store business model and new stores actually to be built. Longstanding organizations tend to cope effectively with disruptive innovation by embracing an extendable core capable of change to their processes and values as needed.

Value Innovation Theory

Successful businesses tend to focus on what the consumer actually wants, but these wants may not be visible to current markets. These hidden wants are considered true value to the consumer and the organization that finds this true value and offers it to the consumer has created a new niche that has not existed until now — a value innovation. Value innovation theory can be used to shape existing innovation ideas that can better deal with value innovation. Comparison of conventional logic versus value innovation logic will explain the idea of this theory.

The conventional approach will accept the current market conditions and plan within those set perimeters. However, the value innovation approach will reshape current market conditions and offer something outside of its set perimeters. The conventional approach will be busy trying to keep up with the competition or staying ahead of it. But, the value innovation approach will pursue fresh innovative ways to create a new market, separating any threat of competition by leaps and bounds. The conventional approach will cater to its current customers through their special, unique needs. On the other hand, the value innovation approach will cater to the overall masses, focusing on their commonalities. This may lose some of the current customers, but should appeal to more of the overall customers. The conventional approach will make decisions based on existing assets and capabilities, while the value innovation approach will make decisions from the perspective of a new business. Finally, the conventional approach will offer to the customer what is determined within the industry's traditional boundaries. The value innovation approach may move beyond the boundaries in search of the customer's total solution.

The value innovation approach involves the ability to assess the current elements of the product or service in relation to its relative level offered to the consumer. This could be displayed visually on a linear chart as a value curve. The dominant market curve must be identified before moving forward with any value innovation initiatives. Success with the value innovation approach depends on the ability to change the value curve by a culmination of adding, deleting or adjusting levels for elements of a product or service.

In 1988, the first-ever movie megaplex built in the world was introduced to the market in Brussels, Belgium. Despite the declining movie industry in Belgium and competition with the new multiplex theatres, the Kinepolis megaplex took 50% of the movie market its first year while expanding the shrinking market by 40%. This extraordinary success was attributed to the value innovation approach Kinepolis pursued by radically changing the current value curve of the competition. Kinepolis eliminated the element of prime downtown location and built the megaplex in a cheaper location further away from downtown.

The new location offered abundant free parking, unlike the limited, costly parking in downtown areas. The cost savings from the cheaper location enabled Kinepolis to offer more amenities to customers, while at the same price of the downtown competition. The 25 movie screens and superior performance provided a superb experience for people.

The companies that have been most successful with the value innovation strategy focused on three areas: product, service and delivery. If the value innovation is focused only on one or two of these areas, over time the returns may not be near as profitable versus the substantial returns from the focus on all three areas.

Both the disruptive innovation and the value innovation theories appear to have the similar goal of finding new markets that will gain a significant advantage over the competition. The disruptive innovation theory focuses on factors of an organization successful with disruptive innovation, however, the value innovation theory focuses on the elements of the current market and determines what could be changed that will result in a new market.

Blue Ocean Theory

The red ocean describes how the current market forces organizations to compete with one other for the limited consumer demands within its boundaries. However, if an organization can create new demand, instead of competing for the existing limited demands, then a new blue ocean market has been created. Blue Ocean Strategy can enable a company to leap into new uncharted territory that offers significant, added value to the consumer and leave the competition behind for a very long time.

This pioneer company journeys into the unfamiliar frontier to find a new product, service or way of doing something that nobody else can copy because they did not travel the same path, the same way, to discover this unique offering that brings new valued satisfaction for the consumer. Any copycats may fail at their attempts to replicate the new value if they are not familiar with the entire unique process involved in achieving the same results with the same acceptable production costs.

Fast Second Theory

Blue Ocean Strategy could be considered a first-mover approach that can be challenged by the fast-second theory, reasoning that a company follows closely behind the first mover within their blue ocean market with the intent to colonize and take over the market that was introduced by the first mover. In a more realistic business atmosphere, it is argued that the blue ocean theory is more or less rhetoric and that red to purplish oceans tend to exist more, at least over the duration.

How Do These Theories Compare?

Disruptive innovation, value innovation and blue ocean theories emphasize the importance of being first to find something new to offer consumers that is unique enough by what is offered and the process involved to offer it that will thwart off any threat of a copycat. However, the value innovation theory focuses on the elements of the current market and determines what values could be changed, added or eliminated that will result in a new market value. The fast second approach is the only approach mentioned that resonates with the advantage of being a copycat if the follower responds rapidly, taking over and dominating the new market. It is argued and a matter of perception whether a market can truly be a blue ocean or more of a gradient of red to purplish conditions. It can also be left to different interpretations whether a new value is actually blue ocean, disruptive, value innovation or more of a vast incremental improvement of an existing market.

CHAPTER FOUR: STRATEGY

You must be relevant to make some money, but all this talk about trends, competitive advantage, dynamic capabilities, and innovation doesn't help much unless you have a strategy to execute all of these elements into actions that will best serve your business needs. A firm's strategy is like a football playbook used for outperforming against their competition on the field, but in a business sense it positions the organization in the market to gain a competitive advantage while maximizing profits, as well. The company will either offer the consumer more value than their competition or the same value, but at less production cost to the organization. If their processes for attracting more consumers are distinct enough from the competition, the organization will be able to repel any competitor's counterattack and sustain a competitive advantage.

The Strategic Plan

An effective strategy will exhibit how an organization will achieve their performance targets, adapt to economic or market changes and seize lucrative growth opportunities. The organization will need to gain the trust of their stakeholders with a business model detailing a plan to provide value for the customer at a desired price and demonstrate a formula that reveals the profit margin for delivery of that value at the said desirable price.

A successful strategy will need to fit the right company within a given industry involved with specific competitive challenges under certain market conditions, while assessing internal capabilities for that company to be able to execute a strategy. The company will need a unique approach to gain a sustainable competitive advantage over the competition that results in strong company performance, often measured by obvious indicators such as increasing financial performance or substantial market share gains. The extent of how well a company performs and gains ground in the market depends not only on the creation of an effective strategy that fits what the company is about, but it also depends on how proficiently the company can execute it.

Leaders must first develop a strategic vision to determine the future direction of an organization and envision their role in that future direction with a justifiable reason that would be in its best interest, unique only to this particular organization. Then they must create a mission statement that shows the organization's current role, what they are doing now, and their reasons for doing it. Leaders will dictate organizational core values that reflect the personality of the company with the expectation of how employees will present themselves while conducting business. They will align these core values with their vision, mission and strategy and establish strategic objectives, which translate the organizations mission and vision into detailed performance goals.

Both financial and strategic objectives will be needed in a balanced combination of these two types of objectives appropriately fitting the particular organizations conditions. The importance of these two types of objectives will align with their efforts to increase both financial performance and substantial market share gains. Managers should use a balanced scorecard for linking financial performance objectives to specific strategic objectives that stem from their organizations business model used to impress their stakeholders. This strategic management process will help successfully execute the intended strategy.

Due to increasing regulations and social voices, successful enterprises not only need to strive for economic sustainability, but environmental and social ones, as well. Viral integration will allow an enterprise to reach across all three of these dimensions rapidly in response to the fast-moving environmental and social demands while trying to gain an economic advantage over competition. There are many definitions, models, and ways to evaluate sustainability across the dimensions. In order to engage viral integration, companies must rely on innovative people who are very motivated to pursue sustainability. These people are design thinkers that possess five traits: empathetic orientation, innovation through integration, optimism, experimentation, collaboration and co-creation. An enterprise will need to seize an opportunity that involves viral integration by delivering a lasting positive impression that is sensitive to the surrounding environment.

Innovation Strategy

Firms are constantly faced with strategic innovation decisions so they will remain relevant to consumers. They must decide whether to create new value for an existing consumer market or to offer the consumer an entirely new market to meet their future expectations derived from significant trends. Managers may be able to acknowledge major trends, but they might not recognize how these trends influence consumer aspirations, attitudes, and behaviors. As a result, their innovation strategies may ignore trends, ultimately offering value proposition to consumers that don't meet current market expectations.

An innovation strategy stipulates how innovation efforts will support the organization's overall business strategy. Leaders make trade-off decisions, choosing the best projects that align innovation priorities and are congruent with overarching organizational functions (Pisano, 2015). A well-articulated innovation strategy impacts an organization's performance of innovation. Strategy begins with innovation goals that are supported by measurable objectives. Goals are broad and abstract in nature, targeting specific areas of action. Objectives are narrow in latitude and more detailed in scope, concentrating on specific purposes. Tactical decisions involve project selection and resource allocation. The majority of new product and innovation projects align in the same direction set forth by established goals. However, there may be exceptions for some more radical advances, but done sparingly. Too many unrelated projects in too many unrelated markets tend to be inefficient and are likely to be unprofitable.

Product Strategy

An organization's product strategy involves some of the most significant decisions that need to be made by managers in the most effective and timely manner. The outcome of their decisions can affect the entire product strategy gamut from cost minimization to value maximization. Managers will need to maintain a close reality with their competition and understand the competitive environment to help them decide when a company launches a product. A key decision for an organization's product strategy will be the market entry.

The order of entry with the introduction of new products into the market will be critical for the product launch timing. Early entry can provide an organization with advantageous market positions and resources. Late entry can be beneficial if an organization can leapfrog earlier entrants with better value proposition to the customer. Even though there appears to be benefits for either early or late entries, greater advantages have most often resulted from being first. The competitive structure of the relevant market will influence an organization's entry strategy.

Managers will need to assess the market structure for the type and strength of available competitive products. The types of competition and their distribution intensity within the market should be determined. Examples of competition types include the following: related-category incumbent, same category incumbent, related-category new entrant, and same-category new entrant. Distribution intensity is the level of a product's availability in a market. The level of intensity is commonly driven by the target market size, pricing and promotion capacity, production capability, and amount of post-purchase support.

Competitive knowledge and intelligence will promote more effective innovation decisions that involve product strategy throughout the product lifecycle from simply an idea to the finished product, the next great novelty. Managers should leverage the following general innovation approaches that consider trends: infuse and augment strategy, combine and transcend strategy, and counteract and transcend strategy.

Infuse and Augment Strategy. The infuse and augment strategy allows an organization to retain most of the features of a current product or service and add new ones that address new consumer needs derived from relevant, but sometimes indirect, subtle trends. This strategy will be the most effective approach when an organization continues to provide meaningful value proposition within their current markets to consumers who appear to fall under the influence of relevant trends.

Microsoft infused its new Windows 10 platform with their current Xbox gaming system. This product augmentation will revolutionize Microsoft's capability to provide a ubiquitous gaming experience to consumers on various devices and at different locations. Microsoft learned from their failed attempt to infuse Windows 8 and they are now moving forward with the new platform to try it again.

Combine and Transcend. An organization will utilize a combine and transcend strategy when they retain current commodity features and add new ones with the intent to create new market opportunities. A business will take into consideration of consumer aspirations, attitudes and behaviors influenced from trends when exploring novel markets, often leading to disruptive innovation.

Fiat is looking to partner with leading companies in navigation systems, media, social media, and entertainment markets to create their own Uconnect platform. This innovative system will provide drivers with communication, entertainment, and navigation features that help maintain their focus on driving. This value proposition addresses the increasing concerns of driving distractions, complementing laws for safer driving.

Counteract and Transcend. An organization will leverage the counteract and reaffirm strategy to create a product or service that reminds consumers about the positive traditional aspects of an existing product/service category. Businesses will emphasize any negative connotations of current trends and show how their product/service can counter such nuances, affecting consumer choices.

The digital age for photos has affected the photo printing market negatively. However, InfoTrends believes they can counteract this negative aspect by educating the public about the intrinsic value of photo prints. InfoTrends emphasizes that these hard copy prints can be passed along from generation to generation, appealing to the emotional worth of these visual treasures. They also contend that the storage of digital photos is not guaranteed to be accessible in the future.

Existing Markets

Disruptive Innovation. New upstart companies may often face challenges by the incumbent market leaders when attempting to enter the more popular market segments. They may indirectly wage attack by initially entering the less popular markets. However, the newly established firms may leverage the incumbent's business model in creating a new model that will cater to the lesser attractive market. This segment will be less challenged by incumbents and allow the upstart firm to grow their initial solutions into superior, disruptive innovation that can eventually threaten with the competitive advantage of incumbents in the more attractive markets.

According to disruptive theory, an incumbent firm may try to imitate incremental innovation. Yet, they may not be willing to imitate the upstart firm's radical innovation due to lack of capabilities or any involved switching costs. Therefore, the incumbent's inability to counter the upstart firm's innovation easily poses the threat of disruptive innovation.

An incumbent firm should embrace a strategy that will be proactive to the possibility of disruptive threats by competitors. The incumbent will periodically need to reevaluate their assumptions about cost and value. They should embrace an innovation strategy that fosters the capability for them to find new ways to cut cost and to find new value that will meet unmet consumer needs. Successful market leaders tend to have a disciplined, systematic process enabling them to sustain a competitive advantage through both incremental and radical approaches.

Most of the time an organization will probably innovate incrementally. According to disruptive theory, an incumbent is most likely to fail when they attempt to disrupt their own market. However, contrary to this theory market conditions may dictate the need for them to prepare to disrupt the market before the competition does so. This preparation will require them to acquire the capability to stimulate, manage, and sustain disruptive innovation.

Microsoft is keeping pace with future innovation in various projects and collaborations. They partnered with Dell to share technology and build upon each other's innovations. Microsoft is building their first innovation center within the U.S. Both their current and upcoming innovations embrace major trends such as the cloud, mobile, streamlined communications, and robust security solutions. Office 365's cloud service is a major breakthrough that brings relevant information and documents to workers. Skype Translator is on the horizon as a dramatic way to remove language barriers around the world. Many new features and innovations have been revealed in Windows 10.

The Free Advantage. Upstart companies that can offer a similar product as the incumbent, but with a free price tag can potentially disrupt the market. This of course will depend if the incumbent can effectively counter with a comparable appeal in a timely manner. The more expedient response will help minimize or diminish any threat to the incumbent's competitive advantage. They may need to derive their own free version through one of the following strategies: up-sell, cross-sell, third-party charges, or bundle packages.

Up-sell offers a free version, while providing a premium version with a price. Cross-sell offers a free product, while providing other complementary products with a price. Third-party charges involve offering free products to consumers who are exposed to advertisements paid by sponsors to cover product costs.

Bundle packages may include free products that accompany paid products or a free product may need to be upgraded with maintenance support at a price. Microsoft has countered free competitor software substitutes with their own bundle package, comprised of various levels of paid office suites, complemented by free software enhancing security and Internet browser needs of customers.

New Markets

First and Fast Second. Blue Ocean strategy is undertaken by firms capable of pioneering a new market with added value that is distinctive from existing markets. Blue Ocean is synonymous with uncharted territory where no customers or competition yet exists for a novelty. The key is whether market domination will eventually be achieved from the first mover or fast second approach. Pioneer firms will take advantage of the first mover approach to establish a strong differentiation that will be very difficult for competitors to overcome. Competitors who can follow as the second leader within the blue ocean can focus on colonization to overtake the first mover eventually.

Breakthrough Framework. There have been successes and failures with both approaches, leaving gaps not explained by either first mover or fast second ideologies. The framework of the four breakthroughs rule may better explain these successes or failures. A breakthrough is defined as something significantly distinctive from the current market norm. This framework proposes the following: technological breakthrough, business model breakthrough, design breakthrough, and process breakthrough. Past success stories have exploited a single breakthrough or have advanced multiple breakthroughs.

The Microsoft research team treats technology transfer as a social process, enabling them to work continually toward innovation breakthroughs. The process is people-centered and is built on relationships, communication, and trust. Innovation happens naturally through an ongoing collaboration between researchers and product teams. Their close-knit communication translated into products and services that make computers more useful, reliable, and fun.

Reshaping Strategy

In many current dynamic competitive landscapes, companies can't keep pace with market changes, missing financial or growth opportunities. John Hagel, John Seely Brown, and Lang Davison discussed how organizations must change their strategy from an adaptation to a reshaping approach. An organization will need to go beyond their enterprise boundaries for this new approach. They can either reshape ecosystems, altering industries and markets or they can participate in another organization's reshaping strategy. The shaper will elaborate a new path that will benefit all who accept the new terms, perceived as the only way for participating organizations to at the very least sustain their relevance, but with the potential to find a more rewarding future.

The perception of lower risks and higher rewards will sell the idea of a reshaping strategy to a critical mass of participants in a prompt, mobile manner. This perception relies on three interrelated components: a shaping view, a shaping platform, and specific shaping acts and assets. Shaping view involves convincing participants to believe financial and growth opportunities are inevitable. Shaping platform is a support system providing leverage for participants to do more with less. Specific shaping acts and assets such as bold sacrifices by the shaper and their intelligible, yet selfless use of assets will persuade apprehensive participants to accept the shaping strategy terms.

Marketing Strategy

Companies have primarily focused on developed markets and minimally addressed or just ignored emerging markets, which may have similar characteristics of developed markets, but don't meet the same standards. However, the mature developed markets in the world have become slow or stagnant in growth. In the past if firms considered emerging markets in their strategy, it has been mainly in a peripheral role. They may need to rethink their view and transcend these developing areas as more of a core requirement in their marketing strategy.

The lack of infrastructure has been one of the primary obstacles for companies considering emerging markets. Consumers within the poorer segments demand value they can afford. They have adopted the lower-end cell phones that may not offer all the premium features. However, these simpler mobile phones will meet the basic requirements needed to perform mobile commerce transactions. Consumers have adopted m-commerce as a means to conduct digital transactions that would otherwise be limited with the lack of infrastructure supporting personal computers.

An organization will need to leverage a digital business model that integrates a new core requirement of m-commerce that is compatible with even the basic cell phones. Unique communication challenges in emerging markets should be considered in a company's marketing strategy. Consumers in these poorer segments may not have the same accessibility to information as in developed markets. Marketing campaigns can target this limited segment via mobile alerts to promote business and opportunities for consumers who own a cell phone. These mobile users can further spread these alert messages through word of mouth with non-mobile consumers, exponentially advertising the products or services of a company within the remote segments. Microsoft could leverage their capabilities to reach into these remote areas in similar fashion.

Case Analysis: Lenovo

In 1984, Lenovo Group was founded by 11 scientific and technical staff members of Chinese Academy of Sciences. They focused primarily in markets related to personal computers. However, this changed with their historic acquisitions of the System X server business from IBM and then Motorola from Google, catapulting them into other technology markets. Lenovo has become a provider of consumer, commercial, and enterprise technology to include personal computers, servers, workstations, storage, smart televisions, and a variety of mobile products, including tablets, smartphones and applications. They have expanded operations beyond the borders of China to become a global multinational corporation now operating in Asia Pacific, Europe, and the Americas. Lenovo owns an immense amount of innovation resources comprised of high-quality research and development, the world's most advanced personal computer technology, and more than 5,000 patents.

Lenovo is not an imaginative company like Apple that radically innovates. Their success lies in incremental innovation and process improvement. Lenovo acquired marginal businesses from IBM and Google and renovated their business models to overcome previous obstacles of inefficiencies. They have implemented the technological breakthrough framework to innovate continually with successful outcomes, significantly contributing to the development of China's information industry. They applied a product strategy, which contributed to effective and timely decisions. Lenovo has remained connected with the close reality of their competition and the competitive environment.

Their research and development laboratories have focused on the miniaturization of sensors, processors, and components, first allowing them to transcend the power of personal computers into the smaller ThinkPad laptop. Lenovo's approach with incremental innovation continued to shrink technology, laptops evolving into mobile tablets and smartphones. Miniaturization continues to advance in what is now considered wearable technology.

Competitive knowledge and intelligence has contributed to more effective innovation decisions that involved product strategy. Lenovo's intellectual property has played a key role for future development of their enterprise technology. They rigorously manage their trade secrets. The interaction between technology and the market has been achieved through long-term strategy and short-term patent strategy. Lenovo's innovation aligns with their strategy to provide customers with the world's best-engineered personal computers.

CHAPTER FIVE:
BUSINESS MODEL

You must be relevant to make some money. While considering the faster pace of change and innovation within a competitive market or recognizing oncoming catastrophe before the competition, business model innovation has become more prominent as a way to remain relevant by increasing the agility and speed of organizations to change how they create value. Such changes can enable an organization to stay off financial crisis, find a new competitive advantage, or even disrupt the market with something new and distinct from their competition. Business model innovation fosters the emergence of new market opportunities and achievement of competitive advantages, becoming more important than product innovation to sustain such advantages.

All industries are subject to the threat of disruptive innovations, especially if markets have weak entries for newcomers to enter with their new threats for incumbent leaders to face. Technology or regulation shifts could affect current market conditions that lead to new players with new rules of the game. Even when there is no threat of new entries, incumbents must recognize when they should change their business model. Such indicators to change their model may include shrinking improvements from their innovations, growing consumer acceptance of their innovations, and diminishing revenue or performance. Many past organizations have made the mistake to squeeze out every penny from existing models and ignoring their need for change.

New business models should maintain customer loyalty, while building barriers to market entry. The intent is for customers to continue buying more than once. Relationship building with the customer through various means such as automatic renewals or product/service/support bundles can help sustain the longevity of customer business. The idea is to find ways that make it harder for customers to abandon the organization for an alternative solution offered by a competitor.

An organization could expand beyond a business model, leveraging a platform model to exploit their distinct patents for other organizations to integrate into their business models. The provider could benefit from interlacing their licensed platforms throughout ecosystems, leading towards sustainment that may not be easily imitated or substituted. The bottom line is to ensure the business or platform model shows clarity for generating revenue. However, timelines for monetization may differ depending on organizational patience and the extent of the challenges.

A new business model should contain mechanisms that enable an organization to reexamine assumptions used for the model. Assumptions should be driven from data gathered from various filtered and unfiltered sources. Diverse perspectives from within the organization and from outside (i.e. competitors or customers) will exhaust the possible scenarios, leading to a more robust business model. Organizations should form a portfolio of business prospects they can experiment with the new model and determine the most likely candidates for future opportunity.

The following three macro-dimensions should be considered when building a business model: Value proposition, value network, and financial configuration. The first macro-dimension considers the following parameters: platform characteristics, offer positioning, platform provisioning, additional services, and resources and competencies. The second macro-dimension considers the following parameters: vertical integration, and customer ownership. The third macro-dimension considers the following parameters: revenue model, and cost model.

Microsoft considered all three of these macro-dimensions when they revised their business model radically to take advantage of current mobile trends, while planning for long-term sustainment. They offer a free Office app for the iPad to view documents, but requires a paid subscription for Office 365 to edit documents on this mobile device. Microsoft has basically transformed their business model from selling products, licenses, and devices to selling continual services and subscriptions.

In the business world, innovation is a process intended to achieve a competitive advantage over competitors. It can improve upon an existing product, service or the organization, itself, with an incremental approach or create something new with a more radical style. Business model innovation fosters the creation of new market opportunities and competitive advantages, becoming more important than product innovation to sustain such advantages.

An ideal innovative business model structure will be described in detail, followed by an example of a successful business model used by a publicly traded organization. Deconstruction of their customer's job will reveal some key determinants for their intent to purchase. The current and future landscape will lead to a proposal of the recommended direction to proceed with purpose of sustainable competitive advantage.

Organization Design Model

Leaders at various organizational levels are increasingly being tasked to reallocate resources and redesigning their companies to align with iterative adjustments in company strategy, which translates to their growing dependency upon organization design models. There is no single ideal organization design model. A leader should choose one he or she knows how to employ and that is flexible enough to be applied along a range of organizational scenarios.

The model enables a leader to determine the organization's value proposition, sources of competitive advantage, activities required to deliver the value proposition, leadership style and culture type needed to achieve the value proposition, organization structure, ways to overcome inherent downsides, and organizational practices essential to reinforcing organizational intent (Beeson, 2014). The rapid execution innovation model may be ideal for an innovative organization faced with the need to respond quickly within a volatile high-tech marketplace.

Rapid Execution Innovation Model

The rapid execution innovation model can enable organizations to achieve breakthrough innovation that will transform their existing business models to meet future consumer expectations. This model is structured to create new value through successive experimentation. This experimental continuum transforms radical ideas into the next big thing. However, innovators should be prepared to be wrong.

Trial and error fosters learning from small failures at critical points of iterative development supporting game changing innovation. The developers and innovators remain engaged throughout the various stages of the supply chain as they shape the final outcome from lessons learned along the way. Improbable prospects can rapidly grow into disruptive realization, achieving a competitive advantage over slower competitors who may simply be staging best practices.

Agile Project Management

The rapid execution innovation model compels a responsive approach with project management. Agile project management provides a more flexible manner for managing project fluidity, while still corresponding with the constraints of scope, time, budget, and resources. Companies can adjust project scopes, frequently re-prioritize, and focus their attention on exploiting the value delivered within the available time and resources instead of wasting their efforts trying to improve the planning and scheduling of projects.

Innovative Business Model

Common Structure

A business model is comprised of two basic parts: design themes and design content. Themes consist of the dominant determinants for value creation. Content defines the detailed activities, connectivity with each other and the assigned personnel to perform these activities. The model will illustrate the collective behaviors involved in this process to propose value at a profit.

Many firms may use the same business model type with generic descriptors, but each one may be customized with a firm's own unique tacit knowledge. A scale model will present a taxonomy describing real world characteristics and a typology describing ideal characteristics. A model can be employed as a laboratory for experimental investigation to build upon empirical evidence supporting current theoretical frameworks or to innovative new ideas.

Common Elements

The business model should contain the following elements: customer value proposition (CVP), profit formula, key resources, and key processes. CVP is determined by the job that the customer needs to be done. The profit formula takes into consideration of revenue, cost, profit contribution and production time for the value to be created. The blend of generic and distinct resources from assets such as people, facilities, equipment, technology, products and brands will generate worth for a particular market segment. An established process will foster repetitive production, increasing an economies of scale to deliver value at a profit. The complex interdependencies of these four elements require cautious changes to any single element since it could ultimately impact the other ones, as well. Firms will need to ascertain their capability to change any of these elements with considerations to expected profit gains and ascribed risks.

Structural Understanding

The explicit understanding of a business model and the source of its success will enable a firm to adapt their model with the necessary adjustments to sustain their competitive advantage. A robust learning system will empower an organization to adapt and learn quickly during any abrupt market shifts in avoidance of losing competitive ground. The two major culprits leading to the demise of competitive advantage have been the inability to adapt quickly to change and the extensive controls required for growing complexities within an organization.

Structural Alignment

Core capabilities and design consistency should align internally and externally throughout the entire business model to build customer value, while capable of expeditious adjustments. The use of strategic information will foster foresight lending to prioritization, while monitoring any discrepancies to be corrected. Analytics enable firms to better understand their customers and enhance delivery of value to them. Sustainable business models require the speed, flexibility and mindset of start-up companies if they are to adapt in timely fashion. They should be empowered to disrupt the market, while leveraging their advantage of established capabilities, resources and assets.

Structural Differentiation

Chris Zook and James Allen emphasized differentiation as the key to competitive advantage, the core of strategy. Greater distinction fosters a bigger advantage. The differentiation and its execution lend more to a firm's performance than the firm's business type. A differentiation map provides a formula to derive various distinct ways for a firm to discern themselves from their competition. Three main clusters, each with five categories provide abundant categorical combinations to achieve distinction, leading to competitive advantage. Industry leaders tend to be the most distinctive and build upon their chosen differences to sustain their competitive advantage over the long term. The American company, Priceline, has postulated their unique blend of value for a two-fold clientele, which has enabled them to become the world's largest online travel service provider.

Case Analysis: NYOP Business Model

Priceline's overall mission is to provide discounted prices to consumers below retail price often referred to as below the price line. Priceline is a one-stop shop website for competitive prices of various commodities such as hotels, flights, cruises, car rentals, and travel insurance. In the year 1998, Priceline first leveraged their patented "Name Your Own Price" (NYOP) business model, beginning their slow start, but eventually a significant journey of magnified growth from then $80 to now over $1,300 per market share. The original version of this business model was leaned down into a more efficient innovative model. This efficient version lowered operating costs, stimulating substantial market share growth. Their registered patent has continued to protect their competitive advantage within the tourism market, while generating additional revenue streams with licensing fees.

Business Model Elements

Customer Value Proposition. Priceline's CVP offers a two-fold network medium for buyers and sellers to engage in online bidding negotiations for a particular commodity. Discounted prices are proposed to buyers for tourism-related commodities such as hotel rooms, airline flights, cruises, rental cars, and travel insurance. Price bids are offered to providers for unused commodities that will convert potential profit loss into gained revenue.

Profit Formula. The profit formula focuses on two main sources of income: wholesale and agency sales revenues. Wholesale revenues are determined from a fixed quantity at a set price negotiated between Priceline and each service provider. The difference between the amount Priceline pays and the amount they charge to the consumer equals to their wholesale proceeds. Agency sales revenue derives from percentage-based commissions charged to the service provider and the customer for each transaction. Extraneous income sources include fees from their patent license, advertisements, and websites. The NYOP model distinguishes itself from the traditional sales model in how it fosters a platform for customers to offer a price for a given commodity and the provider indirectly or Priceline directly decides if they will accept the consumer's offer.

Key Sources. The key sources of Priceline's commodities are the unsold hotel rooms, airline tickets, rental cars and related tourism items for a specified period. They are considered perishable goods with an expected shelf life, driving the need for the completion of sales before they expire and their potential revenue ending with loss of income. The commodity value tends to decrease as it nears the expiration date. Upon expiration the salvage value does not exist like it would for other industry commodities. The majority of providers will be burdened with a surplus during the lower demand seasons, initiating a need to minimize profit losses.

Key Processes. The key processes begin with Priceline and service providers signing an agreement, while they acquire unsold inventory at wholesale prices from the providers. Priceline will act as an intermediary between providers and consumers to determine the commodity's minimal price threshold that the consumer must reach in their bid offers for a commodity. If the consumer's bid exceeds the threshold then a transaction will be completed at the price offer. If a bid does not meet the minimal limit then it is considered unacceptable and the consumer can no longer bid on that respective commodity.

Discovering the Job

Priceline's value proposition is two-fold: providing attractive competitive prices and features to customers and increasing the sale of unsold commodities for providers. They have transformed the travel industry by acting as an intermediary between consumers and providers. A new market segment was exposed with fresh shoppers who could not previously afford to purchase such liberties or were just tight with their money. A new channel has been found to exploit the provider's surplus inventory for additional revenue that would otherwise be recorded profit losses.

Customer Values

The following attributes of the typical Priceline consumers may influence their intent to buy:

- The majority of buyers are sensitive to price, but they are also attracted to innovative offerings.
- The majority are leisure buyers who do not need to buy, especially during economic downturns.
- Premium price consumers prefer better security and more amenities.

- Primary customers have a short attention span and prefer ease of accessibility and use to promptly find and pay discounted prices for commodities that will remain within their budget constraints.
- Secondary customers prefer to compare features with prices.
- Many customers are competitive in nature and enjoy bidding for the best deal.
- Some people are curious to explore new innovations such as bidding a price.
- First-time buyers are attracted to strong recognizable brand images.
- Repeat buyers connect reputable brand images to quality, price or certain market segments.

Provider Values

The following characteristics of the typical Priceline provider may influence their intent to sell:

- Most providers will more than likely have a surplus for at least during lower seasons versus peak seasons.
- Providers will look for cost-effective ways to offload their surplus.
- They will act in their own self-interest to assess viable avenues to eliminate excess inventory.
- Providers prefer to leverage reliable online services, which will realize cost savings that can be passed along to consumers in the form of discounted prices.
- They will avoid methods other than online companies because they tend to be more costly and less effective than online options.
- Providers will monitor trends and forecasts to help them allocate resources.
- They will investigate new markets such as overseas segments to expand and increase earnings.
- Providers look for ways to improve quality customer service to increase profits.

- They prefer cost-effective ways to manage market risks.

Competitive Landscape

Current Conditions

The tourism industry has grown from an exclusive few who travelled in the nineteenth century to becoming a major influence of today's world economy. Natural and man-made disasters gave rise to the challenges for this industry, but people found ways to bounce back from terrorist attacks and tsunami storms. The online element has spawned growth within this industry, attracting more consumers who may not have previously travelled. Online tourism services have matured in the U.S. and European sectors and now eyes have focused on potential embryonic segments. The emerging online travel markets in Asia are small in comparison to other foreign markets, but substantial possibilities lie wait for continued progression.

Future Outlook

The commodities within the tourism industry are not considered by consumers to be necessities. The customer demand will tend to be cyclical in nature. Traveler expectations lean towards quality commodities at discount prices. Globalization, especially within emerging markets of Brazil, China, India and Russia, will become necessary in sustaining a competitive advantage for the future. Each of these distinct cultures will dictate their own unique processes to search, plan and reserve a commodity. Brand integrity will prove ever so critical to gain the trust of these sprouting global consumers. Cultural knowledge with all of their sensitivities will lend to an effective marketing strategy and building customer relationships. Regulatory forces within each country will challenge collaborations of local providers and customers in these budding locales.

The numerous stakeholders involved in the complexities of global tourism will present conflicting goals such as the preservation of nature versus the diffusion of new communications technologies. Limited natural resources such as fossil fuels will force providers to innovate new alternatives. The saturation of services and competition within mature markets will demand exploration of new frontiers, both geographic and disruptive in nature.

CHAPTER SIX:
ETHICAL CONSIDERATIONS

You must be relevant to make some money. Honest and moral companies will gain loyalty with their customers, thus remain relevant. An aggressive milieu can often tempt organizational leaders to ignore, condone or even encourage immoral behavior if forging a competitive edge over rival companies. They rationalize such conduct with their belief that the ends justify the means. The leader will need to avoid such temptations and sew a moral fabric into the inner lining of their organization. Moral leadership and their available mechanisms contribute to the ethical orientation of an organization, impacting their strategy formulation.

Moral Leadership

Effective leaders will prioritize more on opportunities to produce results, instead of resolving problems to avoid organizational detriment. Responsible leaders leverage their strengths and available resources to ensure that strategic goals have been appropriately undertaken. Sincere leaders will take the initiative to pursue a vision they believe in strongly, overflowing an organization with infectious motivation to follow the same path. Arrogance is not an admirable quality and should be stricken from a leader's personality when conducting business. A conceited leader can only discourage employees, thwarting optimal performance (Upadhyay, Upadhyay, & Palo, 2013).

A leader's vision derives from the financial and moral aspects leading to a competitive advantage (Sekerka, Comer, & Godwin, 2014). The humility and intellectual curiosity of a moral leader inspires a vision of growth, embracing integrity, responsibility, compassion and forgiveness towards doing the right thing. Their moral compass can be innate or learned virtue. The leader's sense of morality spawns the rules of conduct within an organization known as ethics. These systemic principles are recognized as acceptable standards of practice for an organizational culture, perhaps including their stakeholders, as well. Many leaders may have fulfilled diminutive formal training in ethical principles, only to meet minimal compliance of standards.

Certain mechanisms enable a moral leader to inject their virtuous values into the veins of an organization. These tools include examples such as code of ethics, ethics training, positive organizational ethics, storytelling, and leader's wisdom.

Code of Ethics

Past scandals such as Enron and current recessionary conditions have amplified the need for regulatory initiatives to restore confidence and integrity within the business world. A code of ethics promotes honest and virtuous conduct in compliance with government guidelines. This is a very powerful tool a leader can leverage towards the institutionalization of ethics throughout an organization. This code will better manage the expectations of stakeholders comprised of management, employees, shareholders, competitors and consumers.

Public accessibility to a company's code of ethics will win back consumer confidence, lending to overall corporate social responsibility. In many circumstances, the chief executive officer (CEO) of a corporation is also the chairman of the board. This CEO duality offers an advantage for a leader to incorporate their moral standards into the fibers of an organization, aligning ethical decisions with overall strategy (García-sánchez, Rodríguez-domínguez, & Gallego-Álvarez, 2013).

Ethics Training

Efforts to integrate structured ethics into an organization will involve continually training management and employees about all facets within the code of conduct. An organization utilizing such educational and awareness channels appears to have a financial advantage and outperform those who are not leveraging this avenue to foster an ethical environment. Training should present various real life scenarios that demonstrate meaning for the learner to adopt the ethical practices more knowingly, empowering them to r implement in the actual business setting. These simulated cases comprise real case studies and fictionalized brainstorm conditions of possible choices a manager or employee could face in the future. This training will better prepare them to stay within the ethical boundaries set forth by the code of conduct, aware of the consequences if they do not. The inclusion of employee input through the use of surveys will help foster a more coherent, collective agreement to do the right thing in any given circumstance (Bradshaw, 2013).

Positive Organizational Ethics

A positive organizational climate can excel beyond creating and sustaining an ethical organizational identity. A living code will exist in a positive organization with the belief that to do the right thing is their only foreseeable choice. This organizational perspective is a virtue fueled by a commitment to character, duty and consequences. Positive and negative emotions control the pursuit of moral fortitude through ethically exigent conditions. Individual and organizational identities will influence this collective ethical strength. A positive ethical network consisting of managers, employees and external stakeholders shares common values and goals with a purpose to cope with crisis situations through sustainable positive organizational ethics.

The field of psychology has been a science focused upon ways to remedy ailments or problems of the human psyche. This pathological approach has focused on the negative, on ways to repair and not on ways to grow or improve. Positive psychology reaches out beyond this scope and focuses more on the virtuous and enthusiastic elements that will contribute to growth or improvement. It is the scientific study of positive emotions, positive individual traits, and positive institutions that empower them to flourish. However, positive psychology does not always involve positive thinking. There may be situations that are better suited for negative thinking as the best approach.

Conventional theories focus more on the negative in examining how to solve a problem, but this may discount imagination and creativity. Appreciative inquiry complements this constraint derived from conventional theories to focus more on the positive. It delves deeper into the fibers of the human spirit to examine more of the creative and imaginative influences leading to social innovation. Appreciative inquiry is the process of discovering what maximizes the potential of people and organizations to accomplish great things that were perceived as impossible or unimaginable to achieve, referred to as positive deviance.

Positive organizational scholarship is the formal educational field of study for the effects of positive psychology on organizations. Focus is on the positive climate within an organization and how positive leadership can offer empathy, compassion and synergistic energies to the people, inspiring them to creatively innovate beyond the familiar and take chances without the fear of punitive administrative actions for failure. Any newfound human potential will in turn enhance the growth of the organization's capabilities and resources.

Within the work environment, an upbeat and optimistic energy from leaders contagiously generate a positive climate. Negative forces and adversity may still exist within the positive climate; however, a more positive approach is used to deal with these challenges while the negative forces move from center stage to more of a distant background setting. When something does not go as planned, blame is not placed. Rather a more positive approach is used to focus on not necessarily solving a problem, but seeking an opportunity to find a solution for the issue at hand.

The positive climate will foster a virtuous atmosphere with a leader who displays behaviors of compassion, forgiveness, and gratitude among the people and within the organization as a whole. According to the broaden-and-build theory, the positive ambience will allow people to broaden their range of thought and see more options, in turn, these new experiences will be stored into their internal reservoir for future reference. This could eventually ignite an upward spiral within an organization to create new organizational capabilities and resources to better deal with challenges and adversities.

Personal traits that contribute to the positive psychology include subjective well-being, optimism, happiness and self-determination. People may subjectively differ on perceptions concerning their well-being. Optimism and happiness could go hand in hand, both fueling a positive and powerful drive to persevere through the most challenging of circumstances. Self-determination derives from the human need to belong, but to be unique while bringing value to the table.

Positive psychology focuses on elements that will foster positive leadership. It will better enable leaders to act less on fixing problems and more on leveraging the talents and creativity of people within the organization. A positive climate helps to nurture leadership development and urge experimentation of what is learned without fear of reprisal for failed attempts. This encouraging environment will invite innovative leadership efforts learned from self-development versus a negative atmosphere that may resist and even sabotage potential progressive changes. This more positive approach will seed the growth of resources and capabilities to achieve what were previously considered impossible or nonexistent productive outcomes.

Positive psychology focuses upon an upbeat methodology and environment that supports the leadership coaching cycle for discovering and integrating new knowledge that can build upon the continual growth of resources and capabilities for a leader and the organization. It provides an optimistic platform conducive to effectual learning of new ideas or the rediscovery of existing ideas that may have been tucked away in the shadows of the mind. Positive psychology could influence the leader to focus on the strengths and growth opportunities during the coaching relationship. A positive climate may increase the probabilities to integrate the relationship permanently, transcending it into a self-coaching lifestyle that will decrease the risks of regression to previous ineffective practices.

Storytelling

The most influential leaders appear to be the best storytellers. They can articulate a story in such a way to massage the emotions of an individual, enlightening them with purpose to achieve a greater ideal. Provocative wit can allure the acceptance of a more desired ethical behavior. Humorous, tactful jokes with concealed meaning can indirectly suggest improvements to people within an organization so not to offend them, but rather present an opportunity for them to change (Auvinen, Lämsä, Sintonen, & Takala, 2013).

Leader's Personal Wisdom

A moral leader filled with personal wisdom, from multiple perspectives, will understand and accept life at a superior level. This empathetic and compassionate leader will more easily touch the hearts of other within the organization, inspiring them to do the right thing during ethically questionable conditions. Their strong ethical orientation will balance the pursuit of organizational goals with the needs of their followers, fermenting an ethical culture within an organization (Zacher, Pearce, Rooney, & Mckenna, 2014).

Management between Strategy and Ethics

Leaders could utilize the framework offered by Peter Singer to manage the relationship between business ethics and strategic management. They would leverage an integrative conceptual model of the strategy-ethics relationship (SER) comprised of bi-polar components, topical themes and spanning themes. The bi-polar spectrum places strategy at one end and ethics at the opposite end. Their overall decisions will consider strategy for productive efficiency, craftsmanship, and exchange, while considering ethical elements such as justice, care, human rights, and safety. The maximization of either strategy or ethics tends to be contradictory in nature, a tug of war from each end of the spectrum. The moral intensity of any given situation will be driven by the magnitude, probability, temporal immediacy, proximity and social consensus of the consequences for a particular decision. When a decision cannot easily be determined whether to favor strategy or ethics, a moral leader will tend to choose ethics over strategy, aligning their choice with organizational governance.

Balanced Scorecard

The structured inclusion of ethics will influence strategy implementation if a moral culture exists from the top down within an organization. Values such as responsibility, sincerity, and integrity will need to align between managers and employees. Individuals sufficiently trained in ethics will exhibit behavior and performance to help complete operational objectives leading to the accomplishment of the overall strategic goal that embraces an established code of ethics. A balanced scorecard can be a very helpful tool to translate strategic virtuous goals into specific moral objectives. This scorecard will establish the performance measurements and targets for each objective, enabling managers to prioritize funds allocations for critical performance drivers and to monitor their strategy at any given moment. This real-time approach can more efficiently maintain a balanced relationship between performance and corporate social responsibility.

Case Analysis: Verizon Communications

In 2013, Fortune magazine recognized Verizon Communications Inc. as the most admired telecommunications company in the world, for the second year in a row and bestowed this prestigious honor three out of the past four years. They earned the best reputation for innovation, social responsibility, quality of management and quality of products and services.

In 2011, Ascent Media shook up the security monitoring industry upon its entrance into this market, however, Verizon with their market capitalization of $106 billion, rapidly followed as a fast-second mover with their added value in the connected home arena, offering consumers innovative and energy-efficient cameras, motion sensors and door and window locks. These two new players stirred up an industrial evolution, penetrating a significant market opportunity to increase their consumer shares.

Verizon had a competitive advantage with its current 93 million customers across the nation to help them leverage value innovation in this new market. Verizon ensured a significant shift in the value curve by offering a bundle package for the homeowner interested in more powerful solutions with extra amenities, integrating this new market niche of home security with added convenience, reaping ultimate savings for the customer. Meanwhile, Verizon is rewarded with recurring revenues from its package deal.

In 2012, *Black Enterprise* magazine listed Verizon as one of the 40 best companies for the seventh consecutive year. *Black Enterprise* lauded Verizon's commitment to diversity in its employee base, senior management representation, board of directors, and supplier diversity. African-Americans represented 20% of employees with many in leadership positions. Verizon presents one of the most diverse corporate boards in the country, comprised of 64% females and other minorities. Verizon extended its reach beyond its walls to do business with minority and female-owned suppliers, spending $3.8 billion. In the same year, the Verizon Reads initiative extended a helping hand of $693,000 to nonprofit organizations who are committed to the advancement of technology, literacy, and education for children, adolescents, and adults.

In 2011, Verizon won the superior alternatives award at the Red Hat and JBoss Innovation Awards competition in recognition for its successful migration from proprietary solutions to open source ones. Verizon was bestowed this honor for successfully implementing a standardized process management system in support of their communications management system that results in a better end product or service for consumers.

CHAPTER SEVEN:
SOCIAL RESPONSIBILITY AND SUSTAINABILITY

You must be relevant to make some money. Socially responsible organizations are perceived by consumers to be good for society and will maintain relevance. As mentioned in the previous chapter, acting in an ethical way involves distinguishing between right and wrong, then making the right choice. However, it is not always easy to create similar hard and fast definitions of good ethical practice. A company must make a competitive return for its shareholders and treat its employees fairly. A company also has wider responsibilities. It should minimize any harm to the environment and work in ways that do not damage the communities in which it operates. This is known as corporate social responsibility.

Social responsibility and effective corporate accountability of health and safety performance have been under review for years. However, it has given rise to more political and media attention in the past year following the factory tragedy in Bangladesh. Societal expectations have prompted government and retailers to search for solutions in preventing recurrences and improving management of the supply chain in general. A more global framework is being addressed with consideration to corporate responsibility in adopting more socially acceptable business practices.

The definition of corporate social responsibility (CSR), or often synonymous with sustainability, has been widely debated for at least fifty years, but continues to evolve and be adapted by businesses and society as a whole. There have been claims in research to reveal at least 37 varying delineations of CSR. A company will exhibit standards of good conduct such as providing safe, reliable products and services to consumers, honoring any warranty obligations, ensuring the fair treatment of their employees, and protecting the natural environment. The typical pattern of CSR involves an organization's strategic emphasis to fulfill the commitment of economic, legal, ethical, and philanthropic responsibilities expected by their stakeholders.

From a managerial perspective on a personal basis, CSR is a strategic-level driven, context-specific organizational responsibility, responsiveness, and result (Triple R) from economic, legal, ethical, social, environmental, and discretionary (primarily philanthropic) perspectives with the intent to stimulate the growth and development of the organization and the community. Responsibility is driven by motivations, responsiveness shows the actions taken, and results reflect the performance outcome of CSR initiatives. Intimate involvement within the local community enables an organization to better understand and meet CSR expectations from a community opinion.

The proposed personal philosophy is based on social exchange theory and stakeholder theory to depict the relationship between CSR and organizational performance. The prior ideology expresses how social change and stability is accomplished through the negotiated exchanges between leadership and employees. The latter illustrates the importance of relationships among leaders, employees, shareholders, customers, and the community at large.

Fortune 500 mainstream corporations have adopted, practiced, and achieved a degree of excellence in socially responsible practices. Business acceptance, global growth, and academic proliferation have been and will continue to be driving forces supporting the consistency, stability and popularity of CSR. Any successful outcome from an organization's acceptance of CSR depends upon the alignment of understanding between leadership and employees.

The degree of ethical leadership behavior influences the ethical behavior of employees during the performance of their duties at work. Moral leaders instill their virtuous values into the organization through guidance, open discussions, formal code of ethics, ethics training, and recognition to reduce confusion about acceptable and unacceptable behavior. They lead by example through ethical behavior, treatment, and decision-making.

An organization's most valuable resource is the people who want to make a difference. Employee engagement with CSR depends largely upon what perceived value they receive from their discretionary citizenship behavior. Employee commitment relies on personal feelings toward CSR and the expected social rewards within the organization. Social and environmental initiatives inform employees about CSR priorities. Social exchange within a company encourages employee participation through assurance that leadership acknowledges a job well done.

Moral complexity, metacognitive ability, moral identity, moral ownership, moral efficacy, and moral courage influence a person's values and goals. Moral complexity is the person's capacity to recognize ethical problems. Metacognitive ability enables an individual to appropriately select, apply, and monitor the ethical reasoning skills necessary to solve the ethical problem. Moral identity involves how ethically a person perceives themselves. Moral ownership happens when a person doesn't ignore ethical problems. Moral efficacy describes how an individual feels they can make a difference. Moral courage denotes a person's desire to ethically act in the face of adversity. A person with a highly developed moral compass tends to exhibit values and goals that align with both the business and personal views of CSR.

Corporate social responsibility (CSR) or sustainability can be seen as an organizational duty to stimulate the development and growth of the company and the surrounding community from economic, legal, ethical, social, environmental, and philanthropic viewpoints. Sustainability creates shareholder value, achieves a competitive advantage, and positively affects critical areas of society and the world in general. Companies look first at the external environment within which they operate and research ways to help overcome significant challenges that demand the resources and competencies it has at its disposal. As a result, organizations translate sustainability challenges into business opportunities making business sense of societal and environmental issues.

There are varying definitions of sustainability as interpreted among different managers and leaders. Their interpretations may or may not include economic, societal, environmental and personal implications. However, the majority recognizes the importance of addressing sustainability and many are already engaging it in some fashion. They agree that economic slumps tend to thwart corporate focus on sustainability, especially if the benefits will be more obvious. Divergent views of sustainability are distinct between leaders who know little about sustainability and those who may be self-proclaimed experts.

While the novice leaders captured the definition of sustainability as preserving the feasibility of a business, expert leaders considered all of the economic, environmental and social factors. Expert leaders will address sustainability among suppliers along their value chain, but novice leaders will not in general. Government legislative is a driving determinant for how novice leaders approach sustainability, but not so much with the expert leaders.

Sustainability dictates the need for organizations to take into consideration the growing public concerns about the environment and society, but only if worthwhile in the long term. Companies that are environmentally and socially conscious tend to generate higher profits and more growth versus less conscious organizations. If an organization is not capable of fulfilling acceptable green standards, they need at least to move towards that direction. Sustainable innovators balance between environmental friendly and less friendly projects and determine an acceptable mix, which generates profit, while complying with legal standards and better satisfying public expectations. However, this is not always feasible. The net present value (NPV) at any given point along the innovation pipeline will drive sustainable innovation decisions (Hannachi, 2015). Organizations forge partnerships with sustainable intent and behavior. However, some critical activities or partnerships may be considered environmentally or socially unfriendly, but will be minimized by their frequency and duration.

The reuse of products and recycling of materials are ways for a green organization to reduce waste. Education and training are ways to foster an ecofriendly culture. Sustainable companies tactfully choose more environmentally conscious partners along the value chain. Shareholders are persuaded that going green will increase their return on investments. Socially responsible decisions support organizational standards for diversity, impartial treatment, fair pay, safety, and health. Strong influential leaders justify philanthropy as a powerful marketing apparatus to win over the hearts of stakeholders. A company with heart, yet financially sound well into the future.

Leaders in Social Responsibility and Sustainability

Mary Parker Follett is considered a management prophet who contributed to the ideas of social entrepreneurship and sustainability. Follett considered a business to be a social institution that could provide a higher purpose for its professionals. Employees were not necessarily driven by their loyalty to a company, but rather the higher purpose of the actual work. The social entrepreneurship model measures the success of an organization not only through the means of profit and return for a business, but also the added value benefiting society. An organization that achieves economic and social goals supports the competitive advantage of sustainability. Organizations influenced by the idea of social entrepreneurship are created by people to serve the people, leading towards the growth of society. Relationships between businesses and governments respond to societal needs of a humanitarian nature such as poverty, impoverished healthcare, down trodden educational systems or human rights issues.

Business visionary Peter Drucker clarified that it is not a matter of social responsibility being profitable to business, instead it is a matter of business being profitable to social responsibility. There is potential business opportunity in every social and global issue. More organizations are finding Drucker's philosophy to be true and more schools are teaching sustainability that involves socially responsible leadership. Leaders will find business opportunities that create sustainable value. Contradictory to prior illusions, there does not have to be a sacrifice of profit to make socially responsible decisions. Rather these decisions will contribute to sustainability.

Ray Anderson owned and was the CEO of Interface, a commercial carpet company, for over 20 years. After reading literature by authors such as Paul Hawken, Ray became more concerned how his company has been destroying the environment by using resources from the earth to create pollutants that further damaged the natural environment in general. He started to reflect upon what kind of a legacy he would be leaving for his family to remember him in generations to come.

Ray decided it was time to change his company business model from a wasteful, consuming process to a zero waste approach that would restore the environment, instead of tearing it down. Ray overcame many barriers along his way towards the new business model. He determined that if his company was to eliminate waste, they would have to reach back through the entire supply chain and change it as well. This entire transformation worked towards innovations and improvements leading to sustainability of economic, environmental and social aspects.

Sustainability across economic, environmental and social dimensions empowered them to promptly respond or adapt to any fast-pace market changes, environmental issues and increasing social demands, while maintaining an economic advantage over their competition.

CSR Strategy

CSR strategy involves an organization's intent to create and implement social projects that offer economic value and a competitive advantage. Leaders determine what elements will influence the success of creating value from social projects. Such elements involve an organization's social positioning and social planning as determined by the external environment, which includes market and nonmarket stakeholders, and the company's internal environment, which includes resources and values.

Leaders take into account the relationship between these elements to link the impact of social projects on company performance.

The relationship between CSR strategy and company performance involves two ideological perspectives. Social exchange theory expresses how leadership and employees negotiate social change and stability. Stakeholder theory emphasizes the importance of relationships among corporate stakeholders such as leaders, employees, shareholders, customers, and the community at large.

Case Analysis: Apple, Inc.

Social Exchange

Steve Jobs, during his tenure as the CEO of Apple, convinced his loyal followers that the company sold more than just products, but rather a vision representing innovation, quality, and superior design. His charismatic leadership and public visibility influenced the direction of the organizational culture. Jobs promoted the distinct attributes of Apple, which did not include much on corporate social responsibility.

However, under new leadership after the Steve Jobs era, the Apple vision adopted more of a CSR strategy beginning with charitable initiatives. Apple matches employee charitable contributions dollar for dollar up to $10,000 a year. Actions such as these mutual donations showed stakeholders that new CSR engagement has arrived within the company. Apple could no longer afford to rely on their past vision without considering CSR. Their current vision is more aligned within the context of today's stakeholder expectations. Apple will need to embrace sustainability if they are to remain competitive with companies such as Samsung.

Stakeholders

Apple identified their most significant stakeholders to be their customers, employees, investors, and the employees of their suppliers and distributors. Stakeholders are taken into consideration when establishing the criteria for choosing the best-fit CSR initiatives. Apple sees their customers as the highest of priorities when planning and implementing CSR strategy. Without customers there is no revenue generated for Apple, whose customers consist of individual and organizational buyers. They have been very successful targeting the mature and middle to upper income level markets based on a one-size-fits-all strategy. Apple provides higher price products in comparison to their competitors, but customer expectations of higher quality and design are constantly met. Apple engages in recycling and responsible sourcing to meet consumer expectations about growing social and environmental concerns. Apple has consistently been a pioneer in product and design innovation with their efficient and eco-friendly processes. Their results have recruited and retained a substantial amount of satisfied customers.

Apple considers employees as the second most important stakeholders. Their skills, capabilities, and opinions are needed for any process improvements or business transformations. Apple acknowledges the necessity of employee contributions to the innovation and development of profitable products. The organization's social initiatives support fair compensation and offers opportunities for career growth among their employees.

Apple's investors are considered a major priority for CSR undertakings. As one of the most profitable companies in the world, Apple's economic initiatives maximize profits for investors. The future outlook for Apple's growth will derive from decisions made by investors not only from an investment perspective, but from a sustainability view, as well. Investors will increasingly pay more attention to the Global Reporting Initiative (GRI), which provides the ranking of CSR standards among organizations such as Apple in over 90 countries. GRI fosters well-informed investors who use caution as they choose companies that meet their CSR expectations.

Employees within Apple's supply chain represent an extension of Apple's CSR reputation. Apple will terminate business relations with external suppliers not adhering to the social initiatives supporting humane working conditions or fair treatment. As of 2014, progress shows 92 percent of their suppliers comply with a workweek rule of no more than 60 hours.

Apple's Trade-Offs

Apple has cut costs and increased profits over the years by extending their supply chain geographically, but not without the trade-off of CSR risks. Such companies cannot be easily monitored to ensure their compliance with Apple's acceptable standards. Some of Apple's suppliers have been exposed by a reputable news team. Clandestine investigations by BBC Panorama revealed inhumane treatment of employees in various Chinese factories producing Apple products or parts.

Covert probes also found illegal mining in Indonesia that is linked to Apple's supply chain. Mining involved child labor in extremely dangerous working conditions. Apple refused to comment about any of these findings, but they iterated that continuous improvement is ongoing and never done. The trade-off of an acceptable CSR strategy will require more stringent monitoring and enforcement of activities at supplier facilities to ensure acceptable CSR standards are met.

Case Analysis: Samsung

Samsung, the company, produces home appliances and mobile telecommunication devices. Their products encompass televisions, monitors, printers, refrigerators, washing machines, smartphones, and mobile tablets. Samsung is comprised of three business divisions: IT and mobile communications, consumer electronics, and device solutions. This company is part of the Samsung group comprised of 220 subsidiaries around the world.

Samsung appears to be a healthy organization that is likely to sustain performance for many years to come. They have shown their ability to align, execute, and renew itself, showing a capacity to continually change. This constant evolution has been Samsung's norm from the beginning, instead of the just the exception. Innovation success can be attributed to their organization design model, organizational processes, their people, essential innovation measurements, and sustainability.

Organization Design Model

Samsung leverages a repeatable model for innovation that they apply consistently in different regions and categories. Innovation has been the driving force for Samsung's growth, fueled by their ownership of over 12,000 patents. However, they have been party to patent litigations that somewhat defray the profitability of patents.

Samsung's innovation strategy aligns with the overall business strategy, enabling their leaders to make trade-off decisions, allowing them to choose the most appropriate practices that align with innovation priorities and congruent to the overarching functions within the organization. Samsung has primarily chosen to extend their business by improving upon current conditions. Samsung focuses on incremental innovation to improve the user experience for existing products. However, they are prepared for major market disruptions through innovation (Brondoni, 2015).

Samsung's business model relies on three business segments: IT and mobile communications (IM), consumer electronics, and device solutions. The IM division is the key revenue generator for Samsung. The strengths of their business model and agile project management have enabled Samsung to launch quality products within a short time span, adapt quickly to changing customer expectations, and penetrate in growing markets with an expansive product portfolio.

Organizational Processes

Samsung is interconnected into the depths of their ecosystem to find the next great way to make life easier or more enjoyable for consumers and business enterprises, especially when it involves their existing products. Samsung continues to generate significant profits by effectively introducing their products in a timely manner, accompanied with satisfactory customer service.

They have sustained a global competitive advantage through the synchronization of their business processes among their locations around the world. The ecosystem has provided

Samsung with valuable data collected from both within the organization and outside among partners or other organizations within or outside of the organization's industry (Parmar, MacKenzie, Cohn, & Gann, 2014).

Success Factors for Innovative People

Samsung leaders conveyed to stakeholders their pursuit of growth and innovation despite the challenges of the global economy. Their transformative technologies and innovation enabled them to identify new opportunities and possibilities. Despite volatile markets, leaders plan to provide new experiences to consumers that exceed the limits of current markets and technology. They plan to offer new technology and new product categories that encourage future core technology. They continually adjust resources to improve efficiency and performance. Leaders reorganized systems to enhance response times and minimize risks. They will accelerate new growth engines in areas of electronic materials and B2B business.

Essential Innovation Measurements

Samsung measures innovation progress by their performance in productivity, sales, and growth. They measure the financial targets for their innovation projects, which have consistently exceeded expectations. Samsung efforts have shown a significant positive impact on profitability and cost cutting efficiencies.

Sustainability

Samsung focuses on sustainable growth to secure their future. Their management practices take into consider the environmental and social responsibilities. They support ecofriendly activities, products, workplaces, and reduce greenhouse gas emissions. They engage in corporate citizenship and social responsibility. Samsung educates and trains their people, while ensuring employees have sufficient healthcare benefits. They are striving to foster positive change for people around the world and improve their lives. Samsung grows alongside communities with shared social values. They forge partnerships with sustainable intent and behavior.

CHAPTER EIGHT:
ORGANIZATIONAL CULTURE

You must be relevant to make some money. The health of an organization will influence their likelihood to sustain performance over time. Their health is derived from the organization's ability to align, execute, and renew itself faster than the competition. They must foster an environment that promotes the capacity to change continually if they want to remain relevant. This constant evolution must become the norm, instead of the exception. An organization will need to focus on not only operational performance, but also innovation as an equal partner. Their culture and climate, competencies, and alliances will significantly contribute to innovative success. These contributors will be discussed in more detailed, followed by the analysis of the most innovative company in the world.

Culture and Climate

A positive climate with genuine, trusting relationships with mutual respect will foster a creative, innovative workplace. Inspirational, intelligent leaders who care about their people will fuel ambition to perform challenging tasks and achieve the impossible. Dreams become synonymous with innovative results.

Power of Leadership

Persuasion. Leadership will need to sell the idea of the innovation vision, mission, and strategy to the hearts and minds of employees. Leaders should express a genuine, sincere desire to change and inspire people of a higher purpose, which is meaningful and real enough to stimulate an enduring commitment from them. Organizational passion can ignite an unstoppable drive towards innovation success. The employees should be convinced that innovation is the only way forward if they are to remain relevant as an organization.

Expectations. Leaders need to manage expectations about innovation participation from their employees. An established process will provide employees with a way for them to be heard and how leadership receives their voice. Innovation participants should expect honest, positive feedback and rewarded if their idea helps generate a new path forward. Even failed ideas should be rewarded if it spawns a successful idea. Creative space should be provided as a place where less controls will allow more freedom for out-of-boundary ideas. However, some control will be necessary to prevent harm and to manage any conflicts that may arise. Energy and enthusiasm will begin to bursts from innovation participants when they know what to expect from their efforts.

Leading by Example. Leaders must be consistent with what they say and what they do. They should be transparent with employees and pass along information that affects them and the organization as a whole. Inform them about organizational goals, the health or status of the organization, and how employees can contribute towards the achievement of goals. Leaders should offer an open door policy to be receptive to employee feedback or suggestions. Leaders should see the organization in different lenses of themselves, the employees, and the customers. Multi-perspectives will nurture the continued growth of a bigger vision, expanding beyond any preset boundaries of limited views.

Organization as a Whole. Leaders should support a matrix organizational structure versus a hierarchical one, entrenched in functional silos. Stovepipe behaviors with an inward-looking mindset grow out of these subcultures that are unproductive for organizational innovation. The matrix approach can break down isolation between functional areas, integrating silos into a project-centric mindset. People who first and foremost see themselves as functional specialists will refocus on the project instead. Their expertise will no longer cause separation within the organization, but will be perceived by members throughout the organization more as tools to get the job done for the organization as a whole.

Collaboration. Leaders need to foster an organizational mindset receptive to not only internal collaborations across the functional departments, but external collaborations with other organizations. They should outsource non-core activities to free up financial resources that can fund more external collaborations. Looking outside the organization can provide access to new ideas, leading technologies, and flexible resources via a global network, not limited to what may have worked in the past.

Creative Atmosphere

Creativity can be argued as a prerequisite for innovation, contributing significantly to generating new ideas that lead to innovation. For people or organizational cultures that do not naturally possess creativity, they will need to stimulate their brains to perceive things in different ways. People heavily rely on personal experience to expand their thought process. Coupled with online consumer surveys, leaders and employees may want to immerse themselves and mimic the role as their customers so they can gain new perspective. They could visit diverse business and company locations. These new firsthand experiences may be vastly different from their past experiences, igniting new concepts to consider within their own organization.

Conflict Management

The creative process within organizations often involves tension, spawned by the various, possibly contradictory thoughts about current reality and the desired future. Positive outcomes from such tension will rely on the ability to massage contradiction and ease tension. Organizational culture and climate need to be taken into consideration when addressing tension. It may be more realistic to focus on the climate versus culture. An organizational culture can be comprised of underlying assumptions and values deeply embedded, often subconsciously. On the other hand, the organizational climate is more visible by observing the patterns of interactions and behaviors displayed.

Conflict of some kind is inevitable within an organization, but not all conflict is necessarily negative and unproductive. Task conflict refers to disagreements about work content. Differing viewpoints, ideas and opinions could promote new ideas. However, it will be important to evaluate employee performance for any negative effects from this type of conflict. Careful facilitation of this conflict could lead to innovative results.

Emotional conflict is driven by anger, frustration or hostility. There is no positive outcome for this behavior and needs to be minimized through intervention if necessary. Formal facilitation may help people behave with more insight and maturity, thus learning to understand and appreciate differences.

Process conflict involves the differing opinions about how a group chooses to approach a task, process or method. Research infers this type of conflict has negatively affected organizations. High performance organizations tend to have moderately high levels of task conflict, but minimal levels of process conflict. The idea for managing any type of conflict is to encourage a workable mix of debate and exchange of differing viewpoints, deflating the negative aspects of conflict

Formal Innovation Process

Organization leaders need a blueprint that shows them how innovation efforts will support the overall business strategy of the organization. This approach will enable leaders to make trade-off decisions, allowing them to choose the most appropriate practices that align with innovation priorities and congruent to the overarching functions within the organization.

There are various terms describing innovation such as blue ocean strategy, disruptive innovation, incremental, breakthrough, continual improvement, and new-growth initiatives. But from a strategic perspective, innovations fall under one of two categories. One category includes innovations that extend current business by improving existing conditions. The other category includes innovations that generate new growth for new markets, often transcending business models. It will be important to define innovations pursued.

The assumptions of innovations should be identified and tested throughout their projects. Innovations should be segmented into smaller groups containing only one significant element. Smaller amounts of information can be more easily learned, while failures can be more efficiently detected in a specific area, allowing quicker adjustments from learned failures.

A viable innovation system will offer a fundamental, organized approach to a few forefront innovation projects that customers need and the organization can deliver. Innovations will be categorized as short and long term growth goals. Reliable and repeatable processes can be set in place with only minimal organizational changes. An assigned executive will shepherd an innovation task force group dedicated to managing this system and related processes. They will reassess innovation projects, revising their categories or eliminating those that are no longer applicable. The innovation group will collect and manage data, continually looking for patterns. Data will come from within the organization and outside among partners or other organizations within or outside of the organization's industry.

Learning Environment

The organizational culture and its employees are the driving forces of innovation. Leaders and employees need to be orientated about innovation and nurtured through career development. The innovation purpose and intent needs to be transparent to everybody within the organization. Training should emphasize the importance and benefits to actively participate in innovative pursuits. Employees will need to learn to adapt their mindsets and learn skills relative with innovation, aligning individual objectives with innovation objectives that align with the overarching organizational goals. The more who are trained, the higher the probability of transcendence towards a cultural acceptance of innovation.

Competencies

The dynamic competitive environment will demand continual improvements, incremental innovation, or radical innovation. The following competencies will be characteristic of a successful innovative organization.

- Enterprising: The initiative to identify problems or opportunities and to seek out improvement through the gathering of information that may appear either relevant or irrelevant to the organization. The ability to generate ideas, identifying the strengths and weaknesses of them, further synthesizing these ideas, and determining their application.

- Dynamic Capability: The capacity to make prompt decisions in systematically solving problems, ceasing opportunities, and quickly adjusting resources as needed.
- Multi-perspective: The ability to see through the different lenses of various internal and external sources, including those across different industries. The engagement in non-work related interests could spawn novel ideas, otherwise overlooked in the workplace.
- Change Management: The situational awareness of current environmental conditions and the fortitude to evaluate future directions and risks based on current and future strengths, weaknesses, opportunities, and threats. Organizations should recognize both the promoters and inhibitors of innovation.
- Learning Organization: Leaders will need procedures in place to consistently monitor their organization's capabilities and scan the changing competitive landscape. They will benefit from an ability to foresee and secure new business opportunities. The greatest challenge may be how an organizational culture will adapt to change. Such change could force people to change jobs, learn new tasks or adjust to new roles. Professional development and specialized training will lend to them learning new skillsets.

Alliances

The level of innovation, whether incremental or disruptive, will be driven by the level and time of effort by an alliance. Based on recombinatory search theory, Corey Phelps provided a better understanding of how local and distant search methods contribute to innovation. Local search methods allow companies to recombine the more familiar knowledge elements into something incrementally new that can be exploited within a short period of time. Distant search methods will take longer and cost more, but could produce a more disruptive type of innovation.

Networks

Alliance networks can foster an environment for companies to converge into combined forces with newfound abilities to achieve new outcomes, otherwise difficult to obtain by a single company. These outcomes could primarily be to create innovation, but profits and opportunities are to be considered, as well. Network structure represents the pattern of relationships among the partners, whereas, the composition of these stakeholders can be characterized by their consistent traits, resources or capabilities. Ultimately, the purpose of an alliance is to create added value, but the level of difficulty may depend on how it is created. A company that intends to only acquire the resources of another firm may not be as challenging as integrating another company's capabilities or even harder, to spawn innovation out of a partnership.

Partner Selection

One of the most fundamental steps prior to building a partnership system is the partner selection. There is no universal model that will apply to all situations, but the process should be as objective as possible. Organizations in an innovation alliance should present a common purpose, burning desire, sophisticated skillsets, and complementary resources. Yang et al based their selection process for innovation alliances on the following basic criteria: cooperative willingness, financial capacity, complementary resources and technological capability. More specific criteria should derive from characteristics of partners that will be needed for alliances.

Alliances composed of partners who are close in proximity within the same competitive market and expertise may more easily adopt the other's tacit knowledge and capabilities.

However, companies may be reluctant to share beneficial knowledge in fear of opportunism by their competitive counterparts. Alliances with partners distant in competitive proximity may not feel as threatened to share valuable knowledge since these partners may possess different, distinct elements from one another. However, these more diverse partners may not understand and effectively integrate such knowledge into their own processes that would contribute to overall alliance capabilities lending to innovation goals.

Detailed legal contracts can formally control the risk of opportunism and increase cooperation to share knowledge. In reference to alliance network structures, densely interconnected partnerships can informally build social capital consisting of mutual trust and reciprocity among partners to also help defray opportunistic inclinations by protecting information that is exchanged. This will benefit alliances consisting of close competitors in the same industry. Such close-knitted relationships can also be beneficial for distant, diverse partners to better understand new knowledge and assimilate it in their own processes.

Case Analysis: Salesforce.com

Cloud computing is built on an infrastructure to deliver software as a service to organizations. The cloud has provided a way for companies to reduce costs of software licensing, investment in hardware, and operating expenses. Organizations are increasingly adopting this disruptive technology as a more efficient way to achieve a competitive advantage.

According to Forbes, Salesforce.com, Inc. was recognized as the most innovative company in the world from 2010 to 2015. This California-based software company provides enterprise cloud computing solutions for both individual consumers and businesses of all sizes and industries. Salesforce.com offers easy-to-use intuitive software supporting customer services, sales, marketing, social media, analytics, development environment, and mobile technologies. Their software can be deployed rapidly and easily integrated into other software applications.

Nearly from the beginning of their inception in 1999, Salesforce.com has taken the innovative path, first with customer relationship management (CRM) software, leading to cloud solutions, and now reaching six cloud service categories. Four of their categories are among the top eight in the world, a phenomenal accomplishment attributed to their innovative focus throughout the lifetime of the firm's existence. No other software company has matched this same level of success in so many categories. Salesforce.com has successfully established a market presence through their focus on research and development. They recently launched next generation platforms that align with future customer expectations in business intelligence, mobile environment, and predictive decision making,, which will enable Salesforce.com to sustain a competitive advantage for the coming years.

Innovation Culture and Climate

Salesforce.com has provided their employees with stellar benefits, challenging projects, and top-notch training. The organization's positive, supportive work atmosphere incubates a commitment to innovation. Employees chat freely within the company's popular internal social media network to collaborate continually with colleagues and share ideas. Their passion has constantly pushed innovation ahead of the organization's competition.

Innovation Competencies

Change management has enabled Salesforce.com to evaluate future paths and assess risks, based on current and future market strengths, weaknesses, opportunities, and threats. They provide a minimal threshold of innovation training for every individual at all levels in the organization. Their professional development and specialized training has contributed to the skillsets needed for maximizing company-wide innovation processes and results. Employees have been free to unleash their creative talents and skills to achieve innovation, sustaining relevance for the future.

Innovation Alliances

Since 2001, Salesforce.com has aggressively engaged in alliances on a regular basis with other innovative companies such as CISCO, Dell, Adobe, Amazon, VMware, and Fujitsu. These successful partnerships continued building capabilities of Salesforce.com. The contributions from these alliance networks have improved their solutions, and created new solutions, while expanding their scope of support among many businesses, across various industries.

Case Analysis: United Continental Holdings

The intense competitive landscape of the global airline industry has been attributed to past crisis such as the aftermath of the global recessions in 2003 and 2008. Airlines such as United and Continental were able to apply lessons learned from the 2003 recession. They reacted quickly for the short term to focus on increasing revenues, decreasing costs and improving efficiency. Outsourcing was a viable way to focus on the road to recovery and for continued growth thereafter. Money would be saved, but risks existed with the compromise of high-quality security and customer service. However, the long-term vision inclined some companies to collaborate, partner or merge with other companies in their best interest of sustainability.

In the year 2010, United and Continental airlines reached an agreement to merge into what is now United Continental Holdings (UCH). By year's end of 2011, all flights between these two entities were branded solely as "United." UCH leverages bilateral and multilateral alliances with various international and regional airlines to enhance travel options for their customers. They also maintain contractual relationships with various regional carriers to support regional jet and turboprop service branded as "United Express."

These strategic alliances and contractual relationships have enabled UCH to provide more frequent flights, while elevating standards of convenience and quality of service to their customers. UCH achieved a competitive advantage over their competition. UCH has now become one of the top air carriers in the world. For the past few years, fuel prices have become the largest and most volatile expense for UCH. If these prices significantly increase, UCH may not be able to pass along this additional cost through higher fares or other fees. They have been pressured to pursue outsourcing options in saving money to offset the looming threat of higher fuel prices.

Outsourcing Strategy

Outsourcing strategy has helped UCH reduce the fixed investment and variable costs associated with internal operations. The decision of which functions to be outsourced will vary among companies based on their internal capabilities. Core capabilities and non-core capabilities that lend significantly to the competitive advantage of a company should remain internal. The industry standard functions that appear to be common among the competition are the potential candidates to be outsourced. The functions that UCH decided to be the likely candidates for outsourcing include information technology (IT), check-in and baggage handling, and customer service.

Information Technology (IT) Services. Business leaders and corporate strategists can become frustrated with IT-related issues due to miscommunications. Many IT functions to support employees are considered common industry practices that hold no real competitive advantage. Outsourcing could leverage external IT experts to improve the communication and service to the organization's leaders, strategists and employees in general. Most companies outsource some part of its IT function. Outsourcing this function can save not only in IT costs, but indirectly affect sales and general administrative expenses.

In 2006, UCH onshored its IT services for a 10-year contract with the American organization, Electronic Data Systems (EDS), which has now merged into a business unit of the Hewlett-Packard Company. They are the most popular IT vendor for the airlines, supporting back-office functions.

The EDS outsourcing agreement upgraded the overall IT infrastructure while reducing operating costs. EDS provides desktop, helpdesk and managed services. They have updated the end-users' computing and communications environments in offices, reservations centers and airport locations. This includes improvement of more 36,000 desktop computers. EDS has provided helpdesk support for employees with their desktops, printers, scanners and networks in over 320 locations worldwide. The vendor has been responsible for any tasks related to asset management, IT procurement and third party providers for IT support. EDS has successfully upgraded IT systems with more innovative technology to deliver high-quality support for all aspects of the airline's computing and communications environments from networks to desktops.

Check-in and Baggage-handling Services. UCH chose vendors to perform many of the check-in and baggage duties across various U.S. airports at lower costs versus their own employees performing the same tasks. These type of tasks generally only required common skills that did not include any distinct capabilities. Initial check-in duties have transitioned to the use of kiosks that minimizes the need for human interaction. Customer service have become important when incidents arise from these type of services that require face-to-face communication in resolving problems or challenges encountered by the customer.

Customer Service Operations. UCH have not maintained customer service standards due to insufficient job training and lack of control over outside contractors. Outsourced personnel have been harder for UCH to manage versus the daily management of their own employees. UCH previously offshored portions of their customer service operations, but were eventually met with unforeseen challenges. Knowledge process outsourcing revealed customer dissatisfaction of customer service representatives who were not easily understood due to language barriers. These obstacles included the limited use of the English language or the pronounced foreign accents when speaking the language.

They reshored these services with onshore vendors to help eliminate these barriers. Another benefit of onshoring in comparison to both nearshoring and offshoring include a cultural understanding to relate better with the customer. The laws and regulations within the U.S. will also help enforce a level of quality work expected by our citizens. Some key initiatives of UCH will focus on additional investment in customer service training and tools for their frontline co-workers to use in conducting business with the customer.

Vendor Selection Process. The primary tasks in outsourcing involved vendor evaluation and selection. The traditional approach for selecting a vendor has relied solely on cost factors. UCH eventually realized this was not sufficient in meeting their needs. Due to the global competitive environment and dynamic customer demands, decisions should not be solely based on cost. Supplier quality has been argued to be one of the most critical factors in choosing a vendor. Their quality of services will affect the accuracy and timeliness of flight and luggage delivery that are crucial in meeting customer expectations. Vendor productivity and flexibility are interrelated and reflect the image of UCH. Their production scope identifies the productivity, process flexibility, and support capacity. UCH replaced vendors if they were not able to adapt to the current challenges presented to UCH.

CHAPTER NINE:
RISK MANAGEMENT

You must be relevant to make some money. Risk can be defined as the possibility for a problem to result from an action or event occurring with person(s) and/or thing(s) at a particular time and/or place. A risk equals to the probability of the occurrence multiplied by the magnitude of the consequence. Depending on the mission of an organization, the consequential aftermath may lead to lost opportunity, financial ruin, or even irrelevance from what could have been avoidable circumstances.

In the year 2011, the conjunction of a magnitude 9 earthquake and a tsunami penetrating land by 4 km triggered an emergency of the Fukushima Daiichi nuclear reactors in Japan. The contributing factors for this nuclear chaos were not considered in the reactor design even though earlier scenarios demonstrated them as possible occurrences.

More than 1,100 people died when a factory building collapsed in Bangladesh. The lack of consideration for risk management procedures and social responsibility contributed to one of the worst industrial accidents in the world. Societal pressures have prompted for government and retailer actions to find solutions to more stringent risk management and socially responsible processes.

Such tragedies could have been avoided or minimized if risk management protocols were in place. Discussions will begin with a concise explanation of risk tolerance, followed by the organizational structure and governance for risk management practices. Some challenges for risk management will be identified. A depiction of the ideal organizational culture and their risk management approach will represent how they can better face these challenges, avoiding another Fukushima Daiichi or Bangladesh incident.

Risk Tolerance

An organization's tolerance for the uncertainty, or their risk appetite, can be driven by strategic choices or the ethical climate. A more benevolent ethical atmosphere fosters a higher propensity to engage with riskier behavior. Risk appetites can vary in different cultures around the world. An Uncertainty Avoidance Index (UAI) can be used to measure a culture's risk appetite. Knowledge of such differences can better prepare a decision maker who has to deal with a business partner from another country. UAI differences between two partners can impact how much risk is tolerated when developing strategy or establishing goals.

Organizational Structure

Risk behavior within a particular organization is influenced by a leader's traits, organizational context and characteristics of the problem or situation. Their disposition will set the pace for an organization's analysis and management of risks. Performance of an organization will rely on the leader's ability to avoid devastating risks when making decisions. The highest levels of leadership will establish the path to be taken with enterprise risk management (ERM) practices throughout the entire organization as a whole.

Every level within the organization should be held accountable and employees should be encouraged to admit to any mistakes without a fear of reprisal. Mistakes are lessons learned, which will lend to a continuous improvement of risk management and add to a knowledge management repository for decision makers to use in future scenarios with similar missteps. Financial and operational metrics could provide measurements for decision makers to consider when determining acceptable risks.

Risk governance establishes a structured process to ensure an organization consistently makes sound decisions based on ERM. Their business model depicts the organization's various risks and provides decision makers with the appropriate responses to these risks. ERM oversees the interdependent relationships across the many categories such as strategic, compliance, operational and financial risks. Internal auditing will assure risk management is governed sufficiently and an independent assessment is provided to leadership that may require them to adjust. The ERM process will identify potential risks, assess the severity and probability of risks, prioritize risks and plan the appropriate response for risks.

Challenges for Risk Management

The identification of risks will either be familiar or unfamiliar to an organization. Managers may avoid newly encountered risks since they may be more difficult for them to manage. They tend to focus on the next risk, which may be more manageable. The assessment of risks will determine the most urgent risks based on their severity and probability of occurrence. In many cases, the assessment may retain and prioritize risks that can accurately be measured for their severity and probability of occurrence, while the more ambiguous ones may be delayed for future consideration or even discarded. An organization would make an epic mistake not preparing for such risks.

Even though management may have identified and assessed risks, the success of risk management will hinge on their responsive actions. Inactions may result in managers who focus too much on the positivity and provide only the answers that stakeholders want to hear without regard to all of the assessed risks. In other instances, managers may prolong any actions of assessed risks until they actually occur. They often believe such risks are fictitious in nature until proven otherwise. Many managers feel that risk management may restrict their pursuance of lucrative ventures or resources are wasted on the investment of ERM. Some managers feel powerless in effectively responding to an assessed risk, any given solutions do not sufficiently address their specific response needed for a particular risk.

Ideal Risk Analysis and Management

Organizational Culture

ERM should represent the whole risk behavior culture of an organization, comprised of a diverse, cross-functional representation. The culture will need to embrace a leadership style that will expect prudent risk management practices, revealing real answers for stakeholders instead of only the positive ones. Managers will need to be empowered with the ability to respond to risks and held accountable their responses. A sufficient allocation of resources should be invested into a proactive ERM system, justified by the value of eliminating risks or efficiently responding them.

Risk Management Approach

A manager should conduct a system analysis driven by prospective benefits, costs and risk. They will perform a risk analysis to identify risks and to measure the probability and magnitude of them. The very low probability of a risk with great consequence could be ruled acceptable depending upon the risk behavior of the organizational culture, related costs and the value gained. On the other hand, a very high probability of a risk with low consequence may also be acceptable. The high probability of a significant risk is not recommended as an acceptable risk.

A realistic estimate for the probability and severity of risks becomes necessary when forecasting computations used for ERM purposes. The more challenging estimates that may be hard to obtain can possibly be derived from evidence found for the driving forces of the risks and the impact. Intuition should be integrated into the ERM process to encourage a mental framework capable of dealing with immeasurable risks. Managers should be aware, at the very least, of the immeasurable risks that present significant implications. Scenario planning can establish ERM responses for various realistic circumstances with unfamiliar risks.

An effective risk analysis will detect patterns, trends and early warning signals indicating reasonable possibilities of occurrence. A reasoned imagination is key for creating the unimaginable scenarios that may reveal the unobvious risks. If a manager derives a scenario that could be equally good or bad, then they should weigh on the side of caution and include this possible situation into the ERM process. The exclusion of such identified risks could be costly oversights in the future.

CHAPTER TEN:
INNOVATION PROCESS

You must be relevant to make some money. The rapid pace of innovation has prompted the need for many organizations to adopt an entrepreneurial spirit if they are to remain relevant in the future. Their atmosphere can become immersed in risk taking, failures, uncertain conditions, and complex competition. This type of culture requires the necessary flexibility to complete experimental feedback loops that will continually evolve at iterative stages of product development until a final product is deemed ready for the customer.

Most incumbent companies pursue incremental innovation, while new upstarts tend to find more success in radical solutions. Incremental innovation is basically continuous improvement that will not sustain a competitive advantage for very long without challenges from competitive alternatives. Radical innovation requires organizations to establish efficient innovation processes, while considering technology and market trends, so they can find a new niche that is more sustainable than a product upgrade.

There is no one-size-fits-all innovation process, but rather various alternatives for certain scenarios driven by the types of innovation projects (Salerno, Gomes, da Silva, Bagno, & Uchôa Freitas, 2015). The selection of incremental innovation projects are based primarily on net present value. Radical projects are approached from the perspective to survive and prosper. Volatile market variables may challenge funding decisions stage by stage.

Along the entire innovation pipeline, a company will focus on product development, and market and sales channel development. Radical innovation comes along with big ideas, but the innovator must avoid spreading resources too thin (Weigel, & Goffin, 2015). The assumptions of innovations are continuously identified and tested throughout the project pipeline. Innovations need to be segmented into smaller, more manageable groups.

Reduced amounts of information are simpler to learn, while failures can be easily detected and promptly adjusted.

A feasible innovation system offers a fundamental, organized way for organizations to deliver innovations that meet customer needs. Innovations will be categorized as short and long term growth goals. Reliable and repeatable processes can be set in place with only minimal organizational changes. An assigned leader will oversee innovation group efforts, reassess innovation projects, or eliminate those no longer relevant (Anthony, Duncan, & Siren, 2014). An ideal process design for innovation management should consider inter-organizational processes, external collaborations, and engagement of customers.

Inter-organizational Processes

There is no single type of innovation process, but rather multiple alternatives that will address specific situations and contingencies. The types of innovation processes are driven by the types of innovation projects. The selection of incremental innovation projects are based primarily on net present value. These projects follow a traditional systematic process and decisions can be more easily verified with customer insights versus radical innovation projects.

Radical projects are approached from the perspective to survive and prosper. Market variables may be unclear at the start of a radical project so funding decisions are addressed stage by stage. Along the entire innovation path, a company will not only focus on product development, but also market and sales channel development. The idea of radical innovation is to think large, but not to spread resources too scarce.

The traditional innovation model is the most commonly used by companies. The traditional process is well structured and manages incremental innovation effectively. However, this model can inhibit radical innovation, especially those in forming markets. Rapid execution innovation may better address radical innovation.

Rapid Execution Innovation

The rapid execution innovation model can enable organizations to achieve breakthrough innovation that will transform their existing business models to meet future consumer expectations. This model involves a process structured to create new value through successive experimentation. Radical ideas can be introduced into an experimental continuum, evolving from thought to execution of the next big thing.

However, participants should be prepared to be wrong. Trial and error fosters learning from small failures at critical points of iterative development supporting game changing innovation. The developers and innovators remain engaged throughout the various stages of the supply chain as they shape the final outcome from lessons learned along the way. The unlikeliest of opportunities can rapidly transcend into disruptive achievement, achieving a competitive advantage over slower competitors who may simply augment best practices of an industry.

Agile Project Management. Traditional project management have been effective with projects with a defined scope, estimated task times, a network of task dependencies, and estimated availability of resources to plan and schedule projects. However, rapid execution innovation projects may lack any or all of these elements, rendering this project management approach as ineffective.

Agile project management steers away from planning the project and provides a flexible managerial approach for project fluidity, while still corresponding with the constraints of scope, time, budget, and resources. Companies can adjust project scopes, frequently re-prioritize, and focus their attention on maximizing the value delivered for the available time and resources instead of wasting their efforts trying to improve the planning and scheduling of projects.

External Collaborations

Project Network

The project network is a strategy to be considered when an organization encounters a knowledge-intensive challenge. This type of network is comprised of core members of a project team who invite non-core participants from their personal networks to contribute towards this group's task at hand.

Historically, project teams have been limited to knowledge gathered from their core members, while personal networks have resolved individual issues. This unique combination of the team's knowledge base and the personal network's problem solving capabilities has set forth the potential to generate more informed decisions.

Project networks could benefit project teams limited in size and ideas. By crossing boundaries, outside contributors would expand an organization's myopic perspective. The small core structure remains accountable, while leveraging the talent pool of many. Advice, information, and feedback from external skillsets and experiences lend to decisions finalized by the core members.

Project teams have successfully contributed to product innovation, customer service, and operational improvement. However, the exponential strength of project networks could have realized even more or better solutions. Project networks could improve the quality of outcomes or propose better solutions that would otherwise not be found by just a project team alone.

Clusters of Innovation

Businesses within the same industry will often form alliances known as industry or business clusters. These clusters involve the geographic concentration of interconnected stakeholders along the supply chain for a particular field or industry. The use industry clusters have increased productivity for participating organizations.

However, this type of cluster is limited with their reliance on the physical proximity of organizations in the same industry, possibly hindering the ability to compete on a global scale. This group is also focused on productivity, not innovation. That could stunt future growth and the sustainment of a competitive advantage.

Clusters of Innovation (COI) can be the solution to hurdle over the barriers of an industry cluster, leading to higher innovation outcomes. COI are individual innovation projects grouped together according to their close proximity. These groupings are vibrant ecosystems built on interconnected relationships among various types of companies, ranging from startups to mature enterprises.

Pools of capital, expertise, talent, and information help foster a creative environment for innovation or technology ventures to evolve at an accelerated rate. This rapid pace is driven by a relentless pursuit of opportunity, staged financing, and short business model cycles. COI will enhance the mobility of these resources, leading to the development of new industries and new ways of doing business.

The individual clusters can be grouped together to form Super-Clusters of Innovation (Super-COI). Global networks can connect clusters around the world, potentially linking unrelated organizations in the unlikeliest of ways. These networks embrace strategies that deliver benefits beyond the limitations of proximity groupings, while achieving efficiencies and innovation on a global scale.

Open Innovation

Mutual cooperation to exchange ideas through an open innovation paradigm can build on the power of networks. An open forum allows firms to leverage both internal and external ideas, creating a partnership to share both the risks and rewards in the pursuit of innovation. Members of an organization that excel in open innovation share a common perspective of who they are and recognize how outside organizations can complement their innovation endeavors. Innovative mindsets will examine new markets based on their competence-based advantages to produce customer value.

Relationship building will tighten the trust and associations among group members, encouraging the willingness to share intellectual and tacit knowledge. Any value-added activities generated from both within or outside the organization should be rewarded with comparable incentives. The open innovation lens will align innovation with strategy, perceived as more than just new product development or marketing.

A cluster of innovation strategy should take into consideration the commercialization of new technologies, creation of new markets, and connectivity with global markets. The three components of universities, government, and entrepreneurs could significantly influence the outcome of clusters. The open collaboration of universities with private industry have helped high-tech firms flourish. Government and military research funded endeavors have founded new ways to do business. A highly educated and technically skilled workforce has often exhibited an entrepreneurial spirit to drive innovation.

Engagement of the Customer

Customer contributions can shape the new product scope and functionalities through iterative stages of development. They could validate the relevance of a new product throughout the entire developmental process, driving any necessary adjustments until the final version is ready for consumption. A company's processes would need to be flexible enough to absorb any changes that could happen at any stage from front end to back end of development. Any potential random changes can be attributed to the fact that consumers may often change their minds or discover new needs during any stage in development. The organization will need to embrace open innovation if it is to integrate consumer input effectively into its innovation strategy.

Customer Roles

Customer roles within fast-paced technology industries tend to be collaborative producers. In slow-paced, nontechnical industries, their roles may appear as a market bridge or innovation promoters for a company. The ability for an organization to access customer insights can be crucial when creating radical new products, services, and business models. More innovation-intensive products demand new approaches, as effective commercialization requires both product and market development.

Market Orientation

Market orientation provides the means for the organization-wide generation, dissemination, and responsiveness to market intelligence. A company will essentially focus on retrieving information about consumer preferences and keeping abreast of their competitive landscape. They will obligate the time to research current trends in a given market and then develop a product strategy that caters to customer needs.

Upon deployment, the company advertises their products as items consumers already want rather than convincing them that the products are something they should want. This strategy is the opposite the approach of past strategies, which would exploit the attractive elements for existing products. Market orientation is more of a collaborative effort between a company and their customers to give them what they actually need without as many compromises.

First responders will benefit the most from market orientation, but sustainability of competitive advantage usually has a threshold of not more than three years. The threshold is driven by consumer preferences and imitable reactions by the competition. Tacit knowledge within the organization will stay off imitation by competitors. Sustainability may not even be achieved if the company's focus is too narrow, inhibiting their ability to anticipate threats from unrecognizable, unorthodox sources.

With more companies engaging in market orientation, the likelihood to sustain a competitive advantage through this strategy diminishes. This approach becomes more of a failure preventer versus a success producer. The adaptability of market orientation can prove invaluable within the increasingly competitive global climate companies now face.

Case Analysis: Sprint

The wireless telecommunications carriers industry appears to be highly concentrated with the four leading companies expected to generate over 93 percent of the industry revenue. These leaders ranked in order of market shares are Verizon, AT&T, T-Mobile, and Sprint Nextel. The market is highly saturated with fierce competition. The rapid rate of technological innovation in this industry creates shorter technology life cycles that force carriers to engage quickly in technology development. The consumer adoption of 4G LTE smartphones is expected to grow robustly in the future. As a lagging company in this industry, the Sprint Corporation will need to improve their innovation process design.

Sprint is currently consolidating and optimizing their spectrums, while modernizing their network platform to support the 4G and LTE technologies that are leveraged by their leading competition. The rapid execution innovation model could enable Sprint to achieve breakthrough innovation that improves the lives of mobile users. This added value could either disrupt the competition or offer a new market all together.

Sprint has acquired Clearwire to continue their foray into 4G mobile broadband technology. This acquisition provides 5,000 additional sites where Sprint can aggressively deploy their 4G LTE technology. The use of project networks and clusters of innovation would further collaborate creative minds to radically innovate better results. The engagement of customers and use of market orientation in these new locations will provide key insight of what the local consumers really want.

Sprint is faced with the challenge of finding new customers in a saturated market, forcing them to pirate customers away from their competitors. Customer engagement and market orientation would enable Sprint to develop new products or features that will attract these customers away from their competitors. Sprint is investing into the expansion of network capacity to keep pace with the outlook of robust mobile usage that exceeds current spectrum capacity. But any spectrum solution within the industry will rely on how much capacity the Federal

Communications Commission (FCC) will allow for use. Their decision could be influenced by effective relationship building that has evolved from within clusters of innovation.

CHAPTER ELEVEN: INNOVATION COOKBOOK

You must be relevant to make some money. Most organizations face technology and market volatility, plagued by a dynamic competitive landscape saturated with new products and services. They must discover new ideas rapidly cultivated into inimitable ways of attracting and keeping customers. Breakthroughs will primarily involve products, services, or processes to either reduce costs or add new value. An innovation cookbook could enable a company to survive the erratic quandary. A summary of what has been discussed in the previous chapters will be encapsulated into a basic mix of guidelines for you to leverage and followed by a case example.

Innovative Organizations

Transformational Style Leadership

Transformational leadership is considered an adaptive leadership style found to be a critical determinant of organizational innovation. Under this style of leadership, essential rudiments from the past will be harvested and recycled into future endeavors, while discarding expendable residuals. Adaptability will foster a thriving atmosphere to promote organizational growth. Transformational leadership is comprised of five principle components: idealized influence, attributive charisma, inspirational motivation, intellectual stimulation, and individualized consideration.

Transformational leaders can identify a needed change. They will leverage transcendental means to facilitate the entire creative progression from their vision of new thought to execution of innovative outcome. These trailblazers enhance the motivation, morale, and job performance of those involved in this process. They challenge followers to perceive themselves not as employees, but partners with vested ownership in their contributions. This type of leader will understand the strengths and weaknesses of workers and align them with tasks that enhance performance results leading to successful innovation.

Innovator Attributes

Disruptive Mindset. Innovation leaders have a disruptive mindset and set the trend for organizational transformation. They have a depth of knowledge in a particular field and breadth in many other areas. Broad experiences enable the innovator to associate unlikely connections creatively, which will lead to disruptive innovation. They remain abreast of customer and industry trends.

Observation and Experimentation. Innovation leaders constantly observe how things can be improved or done differently. They look for competing explanations to challenge their observations. Innovators ask the right probing questions before they can provide answers. They experiment with their products and business models. Innovators remain open to both product and business model innovation. Outcomes from tinkering around can accidentally unravel new ideas. Innovators determine new ideas to be better, faster, or cheaper. Leaders always welcome mistakes. Failures can be the best teacher for innovators.

Networking. Innovation leaders network to discover new ideas. They recognize that face-to-face meetings are most effective in generating new ideas, while the online forums can efficiently capture, filter, and manage the often-large number of existing ideas. Innovators leverage open innovation forums for solving narrow technological problems. Company-specific problems are better resolved internally by people who may have more understanding of a contextual nature.

Innovation Facilitator. Innovation leaders facilitate an innovation project from beginning to end. They reward good ideas primarily by recognition and not through monetary means. Innovation contributors are intrinsically motived by recognition and accomplishment. Innovators follow up with the generation of new ideas, aligning bottom-up and top-down support of an innovation project. They identify both strengths and weaknesses in the innovation value chain, but with more emphasis reinforcing the stronger components.

Innovative Entrepreneur. Innovation leaders emulate entrepreneurs who focus on growing the business. They ensure added value from ideas will contribute to a high multiple growth for the company. Innovation strategically and financially makes sense to them and the organization. Innovators define their company, revealing any inhibitors of new ideas.

Global Perspective. Innovation leaders look through the global lens. They consider economic, political, cultural, and educational factors for an atmosphere before applying a particular leadership style. Highly educated localities are positively related to participative leadership and negatively related to team-oriented leadership. Education levels are directly related with economic and political factors and indirectly to national innovation.

Organizational Attributes

Disruptive Innovation. Michael Raynor described an example of disruptive innovation within the airline industry. Leading airline carriers leveraged the hub-and-spoke system to provide value efficiently to customers, but with tradeoffs not in the best interest of all customers. In this case, the trade-off was the decision for most major carriers to offer flights only at the major airports. This strategy kept their costs low, but created an inconvenience for those customers who may not live close to a major airport. Disruptive innovation involves a business model and technology that enables a company to enter an unpopular market segment and eventually provide new or improved value for mainstream customers that cannot easily be challenged by incumbent industry leaders.

Observation and Experimentation. Southwest leveraged the low-cost carrier (LCC) business model to enter the unpopular segment first, catering to dissatisfied customers who lived far away from major airports. Southwest was able to provide a closer flight for customers who may reside closer to a smaller airport, while realizing profit by minimizing cost. Their lower costs were attributed to the use of one type of aircraft, flying direct, offering only one class service, providing no meals or assigned seats.

These cost savings were not sufficient enough to compete in the mainstream with a higher value for customers. Experimental iterations refined a more cost-effective aircraft enabled Southwest to now enter the mainstream and offer lower prices to customers. This disrupted the industry when incumbent leaders couldn't compete with this added value. The key to Southwest's success was not their strategy, but rather their disruptive business model.

Networking. Shell aligned innovation with strategy to keep innovation innovative. Their initiative for this alignment mainly involved scenario planning and a GameChanger process, which included innovation based on new and emerging technology. Given the various scenarios, Shell would derive the strategic possibilities to explore. Shell shaped its own GameChanger process into an innovation coalition network. They reorganized their portfolio of 85 projects into six interlocking domains.

Each domain contained a group of innovations used in testing and developing a small array of flexible strategic propositions. The domain approach was Shell's way to improve their innovation efforts. They explained how this success was achieved through the Actor Network Theory (ANT). This conceptual framework treats objects as actors in a social network. This theory looks to understand the combinations and interactions of these actors. The domains are considered both actors and actants, which are integral structural elements influencing the entire process to link innovation with strategy. An actant is a binary opposition pairing, such as a hero paired with a villain.

Innovative Business Models

Agile Innovation

Companies rely on knowledge, but challenged by the continually shifting dynamics of how best to leverage it. The agile innovation model provides the basic premise of how an organization can acquire knowledge from anywhere in the world and integrate it into their own operations. There are three levels of knowledge that will dictate how the organization would most effectively engage in this process.

Knowledge Levels. Explicit knowledge is codified, definable, and transferable information that an organization can easily attract through virtual means. Embedded knowledge is context-related and an organization would be required to learn through observation within its initial environment. A small task force would participate in a foraying expedition to the knowledge source location for a temporary period of time.

Existential knowledge is context dependent, forcing the organization to actually learn by hands-on experience within the initial environment for an extended period of time. This learning process would require an organization to build a permanent facility at the knowledge source location, requiring a substantial investment of resources and time. An organization will need to consider what knowledge will work best with their core competencies. Such information should be obtainable, but more importantly have the potential to be integrated into their business structure.

Innovative Processes

Generation of New Ideas

An organization's ability to innovate stems from generating new ideas. Imagination and creativity contribute to new thoughts. From early childhood through adulthood, people have relied on the imagination to generate efficient new ideas without the necessity to rely on experiences (Magid, Sheskin, & Schulz, 2015). Brainstorming has been a way for them to open up new possibilities, but results have usually not met expectations. Failed brainstorming can be attributed to ineffective thought process or ineffective participation.

Brainstorm Constraints. Often the mistake has been when brainstorm participants have been asked to think out of the box with no constraints. Their ideas often lead into improbable directions with no realistic applicability for the organization or participants. They become paralyzed in overwhelming thought. The answer is not to think outside the box, but another box.

Innovation participants should work within a set of loosely based constraints for brainstorm sessions, enabling them to think within a scope compatible with the organization, yet still outside the boundaries of current processes. The box will help control any potential scattered thinking and ease overwhelming feelings that could otherwise stifle the generation of new useful ideas.

Logic Tree. A question would be a way to find the new constraints. The key will be to pose the most suitable question that helps corral the thoughts of participants. A logic tree can provide a systematic process to find effective questions. This process begins with a high-level general question that breaks down into progressive tiers of more specific questions. The best-fit question can usually be found three to six levels down. Any further progression of tiers can lead to a question with too much detail and not abstract enough to produce useful answers in the name of innovation.

Group Dynamics. Many brainstorming sessions have included certain participants for political reasons. But these sessions will need people with original insight relative to the task at hand if the outcome is to meet expectations realistically applicable for the organization. There can be multiple groups at a brainstorming session, but a group should contain no more than four to five participants. This small number of people will force each person to engage in group discussions. All of the dominant participants should be grouped together, allowing the less aggressive personalities to be more open and speak up in the other groups.

Innovative Collaborations

Internal Relations

Proctor and Gamble (P&G) needed to find new ways to discover breakthrough innovations. They pursued four types of innovations: sustaining, commercial, transformational-sustaining, and disruptive. Bruce Brown and Scott Anthony introduced the new-growth factory as P&G's new way to innovate. This factory initially established an educational environment to teach disruptive innovation. However, more emphasis was drawn to transformational-sustaining innovation, finding new benefits in existing products.

P&G started with a two-day workshop with seven new development teams that evolved over time to a full-fledged factory foundation comprised of creation groups and project teams. The factory was initially engrossed with an overwhelming amount of small projects. P&G downsized these projects into fewer, larger initiatives that were more manageable to deploy ideas quickly. This learning environment nurtured innovative mindsets. The organization was restructured to promote new growth. Manuals were created that detailed the processes for the factory's projects. Project demonstrations helped showcase the emerging factory results.

External Relations

Julian Birkinshaw and Peter Robbins examined how GlaxoSmithKline (GSK) Consumer Healthcare addressed their lower-priority brands that would still be lucrative in many of the global sectors. GSK initially created the Future Group to oversee the marketing and Research and Development (R&D) functions for the global brands. The group leveraged an informal creative community to spark innovation, hence referred to as the Spark Network. The network's marketing and research and development environment focused on the continuance of the global brands.

A valuable wealth of new ideas grew out of this network, which also delivered a learning and support system for participants. This network provided an informal, virtual collaboration to generate new ways to grow global brands in an efficient, accelerated manner. It has been an effective way for virtually connecting innovators with one another. This network transcended into a social atmosphere of like-minded people who shared a passion and commitment to generate new ideas, building upon the capabilities of the organization as a whole.

Virtual Communities. Virtual communities have transcended the communication landscape of not only social interaction, but interaction between companies and their customers. An exchange of thoughts about products can ignite an explosion of new ideas between companies and consumers. The outcome of such collaborations will depend upon mutual understanding and the manner of interaction. The concept of Company–Community Interaction Quality (CCIQ) may help explain the collaboration process between companies and virtual communities.

The primary elements contributing to CCIQ include the aspects of activities, sentiment, interaction, and channel. Governance and company capabilities can help measure the quality and execution of activities. Sentiments may not be obvious for evaluation, but they do involve emotions, motivations or attitudes that influence interactions. Interactions between two or more partners, no matter their background, can be observed for frequency and intensity. Those who participate in interactions expect to benefit from either extrinsic or intrinsic reward. The channel refers to the technology infrastructure that the interaction is reliant for virtual communication.

The outcome of CCIQ will influence innovation development decisions regarding the degree of innovation, while considering the maturity, quantity, and quality fit of information received. The degree of innovation will be either incremental or radical in nature. The maturity of information could range from a new to old thought or problem. The quantity of information refers to the amount of needs, new ideas, solutions, prototypes, and product drafts. An organization will need to determine if information is a quality fit. It will need to align with existing company processes or if adjustments before it gives any serious thoughts to integration.

Case Analysis: T-Mobile

T-Mobile provides wireless communication devices and services to consumers. Devices include phones, tablets, and accessories. Services encompass voice, messaging, and data services. They offer postpaid, prepaid, and wholesale options to their customers. Within the wireless telecommunications carriers industry, T-Mobile follows in third place after Verizon and AT&T. This is an intensely competitive and highly concentrated market forcing carriers to keep pace with innovation and technology. Within this saturated industry, T-Mobile will be forced either to compete for existing customers or to find new customers in other areas such as emerging markets.

T-Mobile has paid attention to the increase in mobile phone usage by low-income households, who now represent the majority of mobile users. Mobile subscribers who earn less than $25,000 per year now represent over 61 percent of all mobile users, increasing from 43 percent in 2013. The rise of emerging markets will only add to the need for catering to low-income consumers.

T-Mobile has leveraged a low-cost pricing model to ignite subscriber growth and significantly improve their revenue base. Their lower costs were attributed to business simplifications and operational efficiencies. Cost savings were realized with network optimization, customer roaming, customer service, and improved customer collection rates.

T-Mobile is focusing on alternative customer options such as no required carrier subscriptions, affordable rates for both in country and international use, and extended warranties for consumers who prefer to keep their current devices versus constantly upgrading to newer devices. These niches are leading to mainstream competition that may force Verizon and AT&T to keep pace or risk loss of market share in their industry.

CHAPTER TWELVE:
MOBILE TECHNOLOGY

You must be relevant to make some money. No matter if you work for a business or you are a business owner. Employee success depends upon relevance in the eyes of their employers. Employer success depends upon relevance in the eyes of their customers. Corporations or small businesses are alike in this aspect. And don't be fooled by the term "small business." In 2015, more than 23 million small businesses contributed to over 54% of U.S. annual sales. Many of these small businesses generate bigger revenue than you might think.

Trends tend to dictate relevance. Mobile technology has been one of the most pervasive information technology trends in the past 20 years. From 2012 to 2017, experts project that global mobile data traffic will reach an estimated 11 exabytes per month. Mobile shopping, also known as mobile commerce (m-commerce), has been on the rise as another avenue for consumers to use in addition to e-commerce and actual storefront shopping.

Overview of the Mobile Environment

The advantages of mobile technology and innovations were also accompanied by the disadvantages and challenges both consumers and businesses faced. Debates have risen and perceptions developed from the introduction of new mobile functionality, along with the benefits and hindrances to their use. The prolific popularity of mobile services has driven the surge of mobile device usage during what society is witnessing as the mobile Internet service era. A Cisco Systems study revealed that the number of mobile devices will reach 10 billion by 2017. Within the past few years, consumers have come to expect their smartphones to provide all of their online needs, including entertainment, shopping, networking, traveling, and work. The downside to all of this demand was the amount of data traffic and power required to operate the mobile devices. Mobile traffic was expected to grow 13-fold from 2012 to 2017.

Theoretical Frameworks

The purpose of theory is to speculate how or why a fact, event or phenomena occurred. Naive inquiry may simply develop a hunch without relying on any systemic process. However, scientific inquiry engaged a more detailed investigation of phenomena with a systematic process to prove a speculation as valid or invalid, resulting in new or updated knowledge. Various theoretical frameworks were used by scholars and practitioners to explain the reasons for people's acceptance of new technology related to the mobile environment. Several of the more prevalent theories are discussed.

Theory of Planned Behavior

In the theory of planned behavior (TBP), Icek Ajzen suggested that people's attitudes toward behavior were determined by their accessible beliefs about the behavior. In this theory, an understanding was defined as the subjective probability that the action produced a particular outcome. The evaluation of each outcome contributed to the attitude in direct proportion to the subjective possibility that the person's behavior produced the result in question. This theory can be used to explain consumers' adoption of mobile commerce. The following five factors, listed in order of importance, influenced the adoption of mobile commerce: cost, privacy, efficiency, security, and convenience.

Diffusion of Innovation Theory

Everett Rogers promoted the diffusion of innovations theory to gain momentum into the mainstream scientific community. This theory explains how, why, and at what rate new ideas proliferate through cultures. Diffusion was the means by which innovation was communicated throughout a social system. The following four elements impacted the diffusion process: the innovation idea, communication channels, time, and the social system. The human resources within an organization determined the extent of distribution. Adoption of innovation or technology must be widespread throughout a culture if diffusion was to be sustainable. The rate of adoption relied on the following types of adopters: innovators, early adopters, early majority, late majority, and laggards.

Organizational theorist Geoffrey Moore found that a consumer's attitude toward technology adoption "becomes significant...any time we are introduced to products that require us to change our current mode of behavior or to modify other products and services we rely on." Further, Moore expanded upon the different expectations between early adopters and early majority. He suggested ways to cross the chasm from the early majority to an early adopter.

Theory of Consumption Values

The theory of consumption values provided a paradigm that helped to explain how functional, social, emotional, and epistemic values influenced consumers' behavioral intention to use mobile applications. The increasing demand for mobile apps has been driven by increased availability and options, the increase of business apps used in the mobile environment, and cloud apps development. Also taken into account were new technologies that improved mobile device performance without increasing costs to the consumer. The concept of value took into account the possibility that the consumer purchased a product for its superior performance, economical cost, and the provider's reputation.

Technology Acceptance Model

The recent growth of mobile device usage has led to the rise in innovation products where lifestyle and social interest adaptations became important in decision-making processes. This evolution has redefined innovation to more of a lifestyle enhancement. The technology acceptance model (TAM) investigated factors affecting consumer adoption of new technology. These factors included social value, compatibility, communicability, perceived quality, and observability. The TAM may have been sufficient to examine the intent to adopt innovation, but not for lifestyle innovation products.

An extension of TAM needed to examine post-adoption behavior with the added factor of complexity requiring special skills, a high level of knowledge, and a considerable learning curve. However, TAM has helped service providers determine the factors that have influenced consumers' intent to accept mobile commerce more readily. In turn, this knowledge helped organizations to focus on how they can accumulate more mobile consumer business.

Technology Acceptance Model and Task-Technology Fit Model

Mobile commerce systems allowed consumers to purchase products or services via the Internet without personal computers. This system created a new mobile business model and changed e-commerce paradigms. Organizations within the medical, insurance and real estate industries have adopted innovative techniques such as mobile commerce to improve their competitive advantage. Research of the relationship between mobile commerce techniques and individual performance would provide valuable knowledge to the scientific community.

The integrative use of both the TAM and the task-technology fit (TTF) model provided a more robust explanation of the relationship between mobile commerce techniques and individual performance than each theoretical model by itself. TAM showed how users' attitudes were influenced by factors such as perceived usefulness and perceived ease of use in deciding to accept and use a technology. The TTF model asserted that information technology (IT) had a more positive effect on individual performance. The IT capabilities should have matched the user's task to be performed.

The tool functionality or the utility factor of mobile commerce technology did not appear to affect the factor of perceived ease of use, possibly due to barriers such as download delays for wireless mobile devices. Mobile users tended to work around this obstacle by choosing to transfer data offline to avoid download delays. Barriers were closely related to access issues such as obtaining a service or paying for a service.

The six types of access were physical, cognitive, affective, economic, social, and political. The lack of ease of use and navigation, limitations in bandwidth, cost, hardware and software functionality, and privacy were all potential obstacles to mobile multimedia services. This study was limited to mobile commerce technology adoption within the real estate industry and may not have been generalizable across other industries amenable to mobile commerce.

Unified Theory of Acceptance and Use of Technology Model

The Unified Theory of Acceptance and Use of Technology (UTAUT) model presented the following critical determinants of behavioral intention to use technology: performance expectancy, effort expectancy, and social influences. Facilitating conditions was a determinant used for the use behavior.

Consumer Perspective. The following application of the UTAUT model showed how the determinants were modified when applied to the consumer acceptance and use of mobile shopping technology. Use of the UTAUT model has shown a direct connection between consumers' behavioral intention and their actual use of mobile commerce services. This information helped business determine how best to allocate their resources to consumers. In the context of users adopting mobile shopping technology, this model was modified to include the additional measures of hedonic performance expectancy and consumer anxiety. It also considered the positioning of the facilitating conditions as an influential factor in consumer adoption. The effort expectancy construct was removed because of a conflict with the facilitating conditions construct, and the assumption presents that mobile users already know how to mobile shop. The construct of perceived ease of use was only applied to actual functions and features of the mobile device.

Within the context of mobile shopping, the facilitating conditions construct was an antecedent of utilitarian and hedonic performance and, under examination, was considered more critical than the effort expectancy construct. Performance expectancy was measured with utilitarian and hedonic performance expectancy. Consumer anxiety about mobile shopping acted as a moderator to leverage the causal relationships in the model. Hedonic performance expectancy was determined to be the strongest predictor of intention to use mobile shopping in the modified UTAUT model. Utilitarian performance outlook was ascertained to be the strongest predictor of the original UTAUT model.

The entertainment and experiential functions and features of mobile shopping improved the effectiveness of mobile shopping. They ultimately influenced the consumer adoption of mobile shopping. Retailers and mobile marketers should enrich the hedonic elements of mobile shopping services with emoticons for visual and emotional appeal, multi-dimensional product views, and various product or service presentation options supported by multisensory cues. Businesses that enhanced the consumer shopping experience through utilitarian and hedonic performance improvements retained more mobile shoppers.

Developing technologies such as virtual reality will provide more pleasurable experiences for e-commerce consumers and could eventually do the same for mobile commerce consumers. Virtual reality is a promising technology that pulls the user into a simulated world or imaginary environment full of rich media and high interactivity. The simulated experience is often enhanced with the use of special visual devices such as a mask or a wall-projected room.

This technology has been introduced to entertainment, games, medicine, and education, potentially to expand into virtual reality commerce (VR-commerce) as an alternative to e-commerce. VR-commerce would take place in a virtual reality store. This virtual online experience would take place on a website where the users are immersed into a simulated environment to indulge themselves as if they are physically located in a shopping mall.

Investigation of consumer behaviors towards adoption of VR-commerce can be based on theoretical frameworks such as a modified version of TAM, theory of reasoned action (TRA), and theory of planned behavior (TPB). The stimulus-organic-response

(SOR) model can help integrate traditional marketing and electronic marketing into VR-commerce. The VR-commerce market could grow to billions of dollars in the future. China has taken the first steps to implement VR into e-commerce activities through their development of a 3D virtual platform.

Hedonic motivation, price value, and habit were three constructs added to the UTAUT model, creating the UTAUT2 model. These new constructs were primarily associated with actual use or action, rather than behavioral intention and from a consumer perspective. The hedonic or extrinsic motivation emphasized the importance of utilitarian value. This construct primarily connects with performance expectancy, consistently the strongest predictor of behavioral intention. From a consumer perspective, price is an important factor since they bear the cost of adopting new technology. The context of habit was shown to be a critical factor for predicting use of new technology.

Business Perspective. The UTAUT model was applied towards the research of the acceptance of technology by larger organizations and individual users. This theoretical model was also used for the study of small businesses. Descriptive statistical analysis provided a better understanding of the small businesses' perceptions of mobile commerce with respect to performance expectancy, effort expectancy, and social influences. A summary of correlation and regression analyses tested the relationships and effects among these UTAUT constructs.

Mobile Technology Defined

Mobile Devices

Mobile devices are handheld computers such as smartphones, tablets, e-book readers, and personal digital assistants (PDAs). They usually contain miniature displays screen and enable the user to input data via voice, touchscreen, and keyboard. These devices typically contain an operating system that can run various types of software applications better known as apps. Most of these devices can be connected to the Internet or other devices through the use of wireless connections, Bluetooth technology, and global positioning systems (GPS). Lithium batteries or similar energy sources are used to power mobile devices.

Usage differs between smartphones and tablets. Text messaging is the top activity on smartphones, followed by taking photos, exchanging email, checking the weather, accessing social media sites, conducting Internet searches, playing games, viewing maps and news, and listening to music. The top activity for tablets is Internet searches, followed by exchanging email, accessing social media sites, playing games, checking the weather, viewing news, photos, and video-sharing sites, and purchasing books, video, and other retail items.

Mobile Technology Constraints

The culmination of novel web-based services and the diffusion of advanced mobile devices will drive the success of the mobile web environment. The rise in mobile data traffic may require service providers to determine ways of offloading traffic to Wi-Fi networks or to predict what content the mobile user will request in the future and download it to their mobile devices via a Wi-Fi network. The mobile commerce architecture consists of the client (frontend interface), server (middleware), and database (backend). Instant connectivity and large file sizes continue to challenge mobile activity at any or all points of this architecture.

The short battery lives of mobile devices such as smartphones require users to recharge those batteries before their smartphones shut down. This significant energy conservation issue needs to be remedied with viable solutions. One solution is user cooperation, which actually offers several benefits. The idea is for mobile users to cooperate and use their smartphones as nodes to help each other transmit data or provide battery energy when needed. Other tasks could include data storage and computations.

The surge of mobile applications in vehicular ad hoc networks (VANs) leverages automobiles participating in VANs as mobile nodes to connect with other participating automobiles in transit. The VAN system creates an efficient wide-range network used for safety purposes by police and fire departments, but it will lead to future business opportunities in mobile commerce, automotive industry, and among researchers.

Mobile cloud augmentation is another solution for mobile device computing and power limitations. Much of the computational code used in mobile applications could be offloaded into remote cloud resources. Mobile cloud computing would increase, enhance, and optimize the processing power of mobile devices. The mobile cloud concept retains the user interface for applications on a mobile device, but the applications logic for mobile devices actually resides on a real-time communications network, known as the cloud, where it shares the same applications logic with other user devices reaching back simultaneously.

Mobile Device Governance

The top trend for business enterprises in 2013 has been the growing surge of personal mobile devices and the use of personal clouds within the work environment. This trend has forced organizations to reshape policy to find the most effective way to manage personal devices and clouds. Organizational rules of engagement will be needed to minimize the security risks of personal devices and clouds, which could breach company data and intellectual property.

Mobile devices have enabled workers to increase productivity, decrease cost, and improve efficiencies within their organizations. However, the misuse of mobile devices can expose workers' organizations to security vulnerabilities that can lead to costly consequences. Higher management within organizations must ensure governance for information shared through mobile device usage is implemented as a structured approach to minimize security risks.

Mobile Learning Aid

According to the results of an experimental study of 109 undergraduate students at a southeastern university, people within a learning environment preferred the use of mobile devices, such as iPad tablets, rather than computers. However, users did not learn as effectively when they used these devices as when they used computers. The distracting novelty of mobile devices and their multitasking capabilities could have contributed to inefficient learning. The future of mobile learning will rely on the usability, dependability, functionality, and performance of the mobile site or app. The most acceptable sites or apps will rely on mobility, interaction, artificial intelligence, and technology-based resources such as simulations or games.

Value Fusion

Mobile devices enable firms and consumers to interact, communicate, produce and consume benefits and value in new ways known as value fusion. The results can include customer value (CV), experiential value (EV), customer lifetime value (CLV), and total customer engagement value (CEV). A research gap prevents full understanding of how mobile technologies will contribute to the conventional wisdom of value fusion. A balance will be needed between creating value for the consumer and creating value for the firm.

A synergy between these two values may present an opportunity for creating additional value for the consumer that will in turn create value for the firm. Trade-offs between these two values will need to be considered for obtaining value maximization. Value fusion could become value confusion if the surrounding conditions and phenomena are not sufficiently monitored and controlled.

The roles of any platform owners (including app developers and social media networks) and profits realized from their mutual customer-firm relationships will help shape the value fusion landscape. The relationship between the different types of consumer value and future purchase behaviors will need to be examined in determining the types of value that will increase purchases. Organizations will need to determine the appropriate advertising campaign and mobile technologies for contributing to value fusion creation while minimizing any situation-related contingencies that may cause confusion.

Word-of-Mouth Communication

An empirical study was conducted with mobile users who use ubiquitous decision support systems (UDSS). Survey results from 218 participants revealed their perception that wireless network characteristics, mobile devices, and mobile applications significantly influenced the quality of UDSS. Mobile word-of-mouth was also considered as a major factor contributing to the quality of actual information derived from UDSS. The quality of the UDSS system and the information it provided determined its usefulness to the participants.

The information systems (IS) success model, TAM, and the information adoption model provided the theoretical research framework for examination of the factors influencing consumer participation with mobile word-of-mouth communication. The IS success model identified system and information quality as the key initial antecedents of system use and user satisfaction. This model was adjusted to investigate the mobile business phenomena, which found both system quality and information quality to be significantly related to mobile website usefulness, attitudes toward mobile browsing services, mobile data service usage change, use of mobile technology, and user satisfaction.

TAM posited that perceived usefulness and perceived ease of use determined a user's intention to accept and utilize an information system. The information adoption model highlighted information usefulness as a mediator between information adoption and the influencing processes to explain how people were influenced to adopt advice or recommendations in computer-mediated communication. The system characteristics of wireless networks, mobile devices and mobile applications influenced the system quality that resulted in system usefulness. The word-of-mouth characteristics of localization, immediacy, and customization influenced the information quality that resulted in information usefulness. The consumers' ubiquitous decision-making process influenced business growth, and it has become necessary for companies to understand consumers' motivation to engage in mobile word-of-mouth communications.

Mobile Commerce Landscape

Consumer Perspective

Perceived Value. The consumers' perceived value of mobile commerce has a significant effect on their intention to adopt mobile commerce, and in determining the development and evolution of mobile commerce. Customers want to maximize their value by reducing the costs of monetary means, time, risk, and effort, while gaining benefits that include utility value, entertainment value, and psychological satisfaction. The perceived benefits for consumers included usefulness, enjoyment, and the freedom to connect anywhere with anyone at any time. Consumers' perceived value of mobile commerce was affected most by the first one of these benefits.

Consumer Trust. Consumer trust is a major factor that influences the adoption of any technology. Past adoption theories have emphasized the importance of technology's usefulness and ease of use when considering the acceptance of it; however, additional considerations for adopting mobile commerce require an extension of the previous models used. Mobile commerce contains three entities: technology, mobile network, and consumers.

Researchers used a survey instrument to collect quantitative data from 222 Chinese university students who were believed to be likely to use their mobile devices to perform mobile commerce transactions. Survey results indicated that consumers' internal beliefs about the usefulness and ease of use influenced their intention to adopt mobile commerce.

Consumer trust has been researched as an independent variable in past studies in mobile commerce adoption, but this study considered trust to be an intermediate variable that led to internal perception-based factors as the key determinants of consumer trust.

Personality-based factors and external perception-based factors such as social influence, perceived risk, and perceived cost were found not to impact consumer confidence significantly in the adoption of mobile commerce. Ultimately, research data findings revealed that usefulness and ease of use affected consumer trust profoundly in adopting mobile commerce.

Telephone interviews with 447 Spanish mobile shoppers revealed that the relational variable of trust and the personal consumer variables of involvement and innovativeness were found to be antecedents that influenced the satisfaction of interview participants. Sonia San-Martin and Blanca López-Catalán described how the interviewees were randomly selected from a database of 3,000 mobile phone users who have previously completed at least one mobile commerce transaction.

Trust and loyalty were determined to be the primary drivers towards consumer use of mobile channels. The higher the level of innovativeness among consumers, the more interests and involvement they had with such new technologies as mobile channels. The users' personal variable of impulsiveness was found not to be an antecedent that influenced their satisfaction. M-vendors should encourage customers' trust and involvement with mobile channels to achieve consumers' satisfaction, but should not control impulse shopping, which could prove detrimental to consumers, resulting in their dissatisfaction.

Marketers should concentrate on enhancing mobile store shopping by providing attractive elements that will win over the satisfaction of potential shoppers. An enjoyable shopping encounter should provide a credible, pleasant experience that appears to be useful for customers.

Potential Risks. Information privacy risks involved four main components: data collection inclusion, unauthorized access, unauthorized secondary use, and data accuracy. Two perspectives of mobile commerce were considered with the potential privacy risks encountered by mobile commerce activities. The extension perception described mobile commerce as an extension of e-commerce, and the uniqueness viewpoint described mobile commerce as more than just e-commerce, offering a set of unique features not available in e-commerce. The combination of these two perspectives helped to provide a better understanding of the challenges faced by mobile commerce.

Beyond the risks associated with e-commerce, mobile commerce also risked revealing personal information from mobile devices. This potential vulnerable knowledge may have consisted of a user's location, device serial number, international mobile equipment identity, integrated circuit card identifier, SIM card ID, social relationships, lifestyle preferences, and behavior patterns. A national survey conducted with 278 mobile users within the U.S. indicated that educational level and age correlated with consumers' perception of privacy risks. The consumer perception of risks will influence their intent to adopt mobile commerce.

User Innovativeness. Mobile commerce growth will depend upon consumer innovativeness and the ways they use mobile commerce, rather than the technological innovation itself. The more innovative customers tended to prefer modernization such as a mobile Internet device over an established personal computer. The millennials now dominate the U.S. workforce as the largest generation. Their ages range from 18 to 34 years old. An estimated 73% of millennials completed purchases with their smartphones. They appeared to be more innovative consumers in comparison to the 66% of average shoppers engaged in mobile commerce.

Ubiquitous Commerce. A ubiquitous society has emerged from the proliferation of the use of mobile devices and the advancement in wireless networks. Consumers have potentially enjoyed an enhanced, more convenient, and personalized shopping experience with seamless interaction that ultimately influenced consumer purchase behavior through ubiquitous commerce, also known as u-commerce.

U-commerce provided the theoretical framework addressing wireless, continuous communication and exchange of data, and how network systems impact society and business. U-commerce was the use of mobile networks to support uninterrupted communications anywhere and anytime. The four broad characteristics of u-commerce included ubiquity, uniqueness, unison, and universality.

Ubiquity was the access to information without the constraints of time and space. Uniqueness applied to the characteristics and location of a person or entity. Unison focused on the consistency of information provided regardless of the access point or time. The goal of universality was to eliminate the incompatibilities among the different information systems. The u-commerce characteristics enabled mobile applications to provide mobile users with a convenience, mobility, flexibility, ease of use, faster access to information, and a perceived secure environment.

By the year 2015, the worth of the global mobile phone market has reached $25 billion, an indication of the worldwide diffusion trend of u-commerce. Consumer reliance on their mobile devices to communicate with others and to stay informed has spawned mobile word-of-mouth, which influenced how they now make business decisions.

Business Perspective

Researchers have looked to improve the future of mobile commerce through improved technologies, innovation and understanding the changing dynamics of mobile commerce. Current mobile innovation relies on competitive forces and interaction between device manufacturers and platform developers. Predicting the future landscape for the supply side of the mobile commerce market presents a challenge for researchers. Social media has significantly influenced mobile commerce. More than 50% of mobile users accessed social media sites, potentially leading to social commerce. This social trend will continue to mesh with mobile commerce in the same direction to become the next new era of mobile commerce.

Consumer Expectations. Many organizations still depend on a traditional physical business model that relies on human interface in meeting customer expectations, but expectations are changing. Customers now demand more capability to interact with companies at anytime and anywhere with their mobile devices. Their expectations may exceed the reality of what firms may be able to offer, especially with security and privacy protocols. The MIT Center for Information Systems Research conducted a two-year research study with 118 companies. The outcome revealed the majority of establishments would replace traditional online channels with mobile apps if available.

However, their preference would be contingent upon the practicality of the product being provided to the consumer. Customers tend to benefit more from mobile commerce with products of high time criticality and low information intensity. Shoppers will use mobile commerce for products with characteristics that fit the mobile commerce capabilities better than e-commerce capabilities. In general, mobile commerce will not realistically replace e-commerce, but it will enhance overall business with increased sales.

Digital Business Model. Organizations may need to transition from a traditional business model to a digital business model or they will risk losing mobile consumers. The digital business model consisted of three components: content, customer experience, and platform. The content described what customers consumed, and included digital products such as movies, software, and information about the products sold or brokered. The customer experience described the digital customer experience with an organization's website or digital business processes for tasks such as shopping carts, payment options, or messaging. The platform involved a coherent set of digitized business processes, data, and infrastructure with both internal and external components.

Internal components included customer data and business processes that did not interact with consumers such as analytics, human resources, finance, and merchandising. External components included the phones, tablets, or computers that consumers used to research and purchase the products, along with telecommunications networks and partnerships with delivery companies. The platform can deliver digital content to the customer and manage physical product delivery to them. The move from place to space and the growing requirement for a digital business model are phenomena for both the consumer market and business-to-business enterprises.

Mobile Strategy. More travelers are using their mobile devices to access travel-related information; however, they tend to become frustrated with slow loading times and how the information is displayed on the smaller screens versus larger computer screens. This growing popularity is driven by capabilities that deliver content to the user based on his or her current location. This consumer trend forces destination management organizations to adopt mobile web technology.

It will be significant for organizations to provide mobile users with an efficient, user-friendly experience that is pleasing to their eyes on smaller screens. The organizational leadership will need to review strategy that will take into account mobile web technology. They will need to determine if they can only adjust their current website infrastructures to more mobile-responsive websites or if they need to develop entirely new systems.

Organizations should focus on mobile web strategies that will exploit consumer emotions of pleasure and arousal to help build relationships between innovation and the consumers. Effective communication between humans should be considered in this equation. Mobile commerce appealed to many consumers with the convenience of time and flexible locations. On the downside, the mobile Internet experience was bombarded by the limitations of the mobile phone, mobile telecommunication networks, and interface problems. The overall consumer satisfaction has apparently been acceptable, but organizations should continue improving the mobile user experience. Improvements should encompass more media rich information, convenient input methods or more efficient infrastructure construction such as wireless terminal equipment.

When considering mobile strategy, a company must compare the use of mobile web versus mobile apps. The mobile web approach has been sufficient for the average consumer; further, the mobile web is cheaper because it will not require individual development that would be necessary with the web applications approach. Technological changes such as a mobile operating system will not affect the mobile web but will affect mobile applications.

Whether firms use mobile websites or apps, they will still need to address the limited storage and computing power of mobile devices. Companies will be required to manage media efficiently and effectively cache data onto mobile devices. They should employ longer data-oriented operations with fewer connections than required of traditional desktop computers. Even with these precautions, a mobile website or app may not be sufficient for every e-commerce service offered by an organization. A mobile approach can provide the user-friendly services for the limited screen sizes of smartphones and their slow networks. Clickable links can be displayed for fixed e-commerce services that will need to be managed at traditional websites.

Mobile Websites. A mobile-responsive web design provides a way to code websites so they will appear adequately on any screen. Instructions can be added to the meta-tag code in the HTML file, which will dictate how an Internet browser is to behave with a particular mobile device. Style sheets that will determine the width and height of the browser window can be added to the HTML files for the respective mobile device display windows. The website navigation menus can be one of the layout challenges for the mobile website design.

An academic health sciences library website was transitioned into a responsive website, and the conversion tripled the library's mobile users within the first year. Responsive website designs are becoming the best-practice standard for developing mobile-optimized websites. They accommodate multiple screen sizes and require less maintenance than traditional sites. Small businesses that develop web-responsive websites will support the various mobile devices efficiently. However, consumer concerns about security, privacy, and utilitarian constraints may still exist, which may be better addressed with the development of mobile apps.

Mobile Apps. Mobile apps are software programs developed for use on specific mobile devices such as smartphones and tablets. The increasing demand for mobile apps has been driven by increased availability and options, the increase of business apps used in the mobile environment, and the development of cloud apps. Also taken into account are new technologies that improved mobile device performance without increasing costs to consumers.

The app developers, who were often highly educated, may have lacked the specialized expertise necessary for app development. Many developers tended to overcome their technical deficiencies with software engineering techniques such as reusing software code. This approach repackaged existing software inheritance, classes, and libraries. The reuse of code has improved development of apps, leading to a faster production time to market and higher quality software.

Potential vulnerabilities exist with the reuse of apps, thus leading to bugs that affect the quality and reliability of the product. Worse, developers may engage in unscrupulous acts of theft or fraud. Other challenges include customized mobile functionality, user access and usability, user interactivity on mobile devices, bandwidth limitations, and unreliable wireless networks. The quality of a mobile app will be essential for consumers to remain loyal and continue using the app. Businesses should ensure their apps contain source codes that do not include repetitive instructions. This redundant type of code will slow down the processing power of the app, which will be perceived as a low-quality app by consumers.

Mobile apps have changed the usefulness of mobile devices from mere connectivity gadgets to lifestyle tools that consumers use to find stores, research products, make purchases, and manage accounts. Mobile phones offer both basic voice service and value added services (VAS). VAS may offer non-voice services such as caller ring-back tones, radio, web browsing, social media, mobile apps, and digital content downloads.

The VAS industry has evolved from simple interactions between users to mobile television and mobile commerce. Mobile commerce includes all activities containing a commercial transaction conducted through communication networks that interface with wireless devices. Examples of mobile commerce include mobile banking, mobile ticketing, mobile coupons, and mobile purchases of goods and services.

Mobile Device Platforms. Mobile apps can commonly be downloaded onto mobile devices from online app stores such as Google Play, the Apple App Store, and the Windows Phone Marketplace. By the year 2016, revenue from the mobile apps market is projected to exceed $22 billion. The majority of mobile consumers use either iOS or Android platforms, which seems to indicate that developers can focus mainly on these two systems when developing mobile apps. Businesses have contained the costs and complexities of leveraging apps more efficiently when constrained only to these two platforms.

The unique elements of mobile device models or versions presented the problem of device fragmentation, which made it challenging to develop a holistic mobile app that would cater to the broad mobile audience. The concern with time has prompted the faster strategy to develop apps first by implementing code generators for the most prevalent mobile platforms, followed by extending to other software environments. The creation of a single mobile device platform to be used on all mobile devices has been a high hurdle to overcome.

The device-independent mobile application generation (DIMAG) framework proposed a solution to overcome this hurdle. The framework defined native device-independent mobile apps based on app workflow, data synchronization, user interface, and data queries. DIMAG considered new devices and new target platforms, while taking advantage of current standards. This new system has been tested on three different platforms: Android, Java ME, and Windows Mobile. Further tests are projected with the iOS platform to validate the design of the code generation module. Analysis of the scalability and the alignment of user interface definition with W3C will augment future efforts for DIMAG.

Mobile Transactions. The ubiquitous computing advantage of mobile phones provides users with the convenience, independence, mobility, and flexibility that can increase their productivity and improve time management. The convergence of mobile devices, wireless communications, and the Internet offers a range of advanced mobile data services to include mobile commerce and m-payment.

The overall adoption of mobile commerce by consumers and merchants will primarily rely on a secure, reliable, and easy way to use the m-payment system that allows for the convenient completion of transactions. Similar to other payment systems, m-payment systems involve banks, financial institutions, consumers, and merchants. Unlike other payment systems, m-payment systems include additional stakeholders such as mobile network operators, mobile device manufacturers, content developers, and content providers.

The success of m-payment systems depends upon the mobile device owner's initiative to use this new innovative channel. The technological advancement of smartphones and apps such as Apple Pay and Google Wallet should build trust and influence consumer use. M-payment stakeholders such as banks do not necessarily hold the same power as with other payment systems. The larger size of mobile customers at such institutions can lead to more complex and challenging negotiations about customer data, cost, risk, and revenue that can influence the sustainability of m-payments. During the past few years, M-payment systems have proven successful in Asia. Citigroup forayed with startup companies in the Asia-Pacific region to leverage their disruptive ideas and technologies of m-payment solutions.

A micropayment system provides an efficient means to perform low-value and high-volume transactions but provides minimal security protection. Target stores suffered security breaches from a similar system with minimal protection. The Smart-M3 micropayment system is a multi-vendor, multi-device and multi-platform infrastructure that utilizes portable and modular technology. Device and service manufacturers who use Smart-M3 may opt to allow their information to be used by third-party services and devices. Third-party usage may lead to the creation of new apps.

A Smart-M4 micropayment system pilot was implemented for use by small to medium-sized businesses with limited resources and capabilities. This method was appropriate for companies that had billing systems. Business users expect this system to provide them cost savings, resource efficiency, and the ability to upgrade the system with future devices or systems. Customers expect this system to improve their user experience to include safety, effectiveness and ease of use.

Mobile Enabled Supply Chain Systems. A model was created to identify what influences a business's intent to adopt mobile-enabled supply chain systems (mSCM) in the retail industry. Supply chain awareness, institutional pressures, and long-term relationships with partners and top leadership support were found to be the principal influences. This model provided beneficial information to help supply chain members overcome innovation barriers more efficiently.

Technical reasons did not seem to influence mSCM adoption, but mobile technology has presented new opportunities for multichannel distribution sales. Mobile commerce continues to give rise to the online shopping experience that is reinforced by more information and satisfaction. It will be challenging for an organization to incorporate mSCM; however, this change would positively affect everybody in the organization and its affiliated business partners. These partners may be concerned about privacy and security issues. A solution will help build their trust, leading to acceptance of mSCM.

Mobile Marketing. Advertising that caters to mobile devices is referred to as mobile marketing. Successful use of mobile commerce relies on the implementation of effective mobile marketing campaigns. In 2011, with 1.2 billion mobile web users, it has become necessary for retail marketers to leverage mobile marketing strategies to meet the surge of mobile commerce. The design of websites typically contains visual imagery and textual content to inform and persuade consumers. Multimedia product viewing, informative content, product promotions, and consumer-led interactivity are significant to an online marketing strategy, whether on the web or in mobile web environments.

Researchers have yet to determine if the better mobile web strategy is to optimize a mobile website or to develop a mobile web application. Mobile apps have grown into a major sales channel that has appealed to many consumers. The design elements of a mobile website or app will determine how effective it will be in encouraging consumer emotions and behaviors that will benefit business. The ubiquitous nature of mobile users requires firms to include a mobile marketing strategy if they are to sustain a global competitive advantage within the mobile commerce community.

Based on traditional technology models, the majority of the factors that influenced consumer perception and acceptance of mobile commerce included usefulness, ease of use, trust, cost, and privacy. According to survey results of 160 Jordanian mobile phone users, the transition towards 4G networks and a mobile marketing campaign strategy focused on consumers who own mobile devices should help improve consumer acceptance of mobile commerce. All of the survey items were measured as reliable for internal consistency and determined as valid.

Active mobile advertising strategies will be crucial to success in this new era of mobile commerce. Advertisers will need to determine how to approach consumers with useful mobile advertisements. The determinants of consumer attitudes towards mobile technologies and for advertising are different. According to proponents of technology acceptance theories, perceived usefulness and ease of use are technology-based external motivators. According to proponents of motivation theories, entertainment and irritation are emotionally based internal motivators. A combination of the model of technology and emotion-based evaluations was integrated into a robust framework capable of predicting consumer response to mobile ads with superior accuracy. This integrated model of mobile advertising approached attitude formation with both technology-based evaluations and emotions based evaluations.

Beliefs about the mobile technologies were formed by the utilitarian considerations of system usage, and attitudes toward mobile advertising were influenced by the hedonic factors of the mobile ads. Advertisers should take into account the level of consumer experience in using mobile services. The less experienced customer was more likely to be influenced by the hedonic content of the ad. The more experienced consumer observed the hedonic content, along with the mobile technology and service used to deliver the ad. Advertisers may need to develop a distinct mobile advertising design catering for the different levels of mobile service experience.

Digital displays are replacing the public static signs as the next generation advertising strategy. They are creating opportunities for interaction with people who choose to engage with digital displays through the use of their mobile devices. The user experience for these types of displays consists of a three-stage process: attraction, interaction, and conation. The attraction stage is the design method used to entice users to the interactive display advertisement and to prepare them for interaction. The interaction stage begins when the user associates with the display to explore the available information and ends when the user has finished acquiring the desired information.

The desired message strategy for the advertisement will determine the stage design for the interaction discreteness and interaction range. The interactive modalities: presence, remote control, gesture, and direct touch determine discreteness and describe how publicly or explicitly a user must operate during the interaction. The range defines how close or far the user must be to interact with the display. The conation stage occurs after the user completes the interaction and chooses to perform an action that has been solicited by the display interaction. These activities may include classification, voting, rating, content control, and content download.

Mobile Branding. An extensive literature review has revealed that the design elements of a website have been found to induce consumer purchase intentions and loyalty by providing customer satisfaction and entertainment value while building their trust. The consumer perception of a website brand is critical to the success of an organization. The growth of smartphone usage now presents a concern for businesses to design a mobile environment that will most likely appeal to mobile users. Since 2011, mobile browser, mobile-optimized websites, and mobile apps have collectively encouraged consumer spending of £438m

($704,203,260 in US dollars) in the UK. The mobile browser can transfer the preferred design elements from a traditional website to a mobile-friendly environment. Branding design rudiments may appeal to website users differently than for mobile users.

Branding literature has identified a sporadic branding of design elements, but these elements have not been organized into a holistic framework to show the various findings for both websites and mobile environments. A framework of mobile branding (m-branding) design suggested four key brand categories: brand name, brand logo, brand design, and brand content. The shapes and icons, along with the colors, typefaces and overall appearance of the mobile channel will create stimuli for consumers to feel positive toward the store.

Consumer Behavior. The increasing popularity of telecommunication technologies, mobile devices, and wireless applications prompted for the important role of discovering user behavior patterns within mobile commerce environments. Consumers have adopted widespread use of telecommunication technologies, mobile devices, and wireless software application that lead to mobile commerce activities. Businesses must examine the patterns of consumer user behavior related to mobile commerce. Their examination is essential for broad applications such as planning physical shopping sites, maintaining e-commerce on mobile devices and managing online shopping websites.

Mobile sequential pattern mining considered users' movement paths and purchase histories to reveal the complete set of sequential patterns within a mobile commerce environment. Interesting mobile sequential pattern mining (IM-Span) ruminated user-specified importance constraints. These limitations may have filtered out irrelevant information that reduced costs for organizations analyzing patterns. IM-Span employed three strategies: utility shrinking, functionality accumulation and progressive match. Experimental evaluations have shown that these strategies significantly reduced the scope of mining and decreased runtime by over 95% on average, while outperforming state-of-the-art high utility mobile sequential pattern mining algorithms. The IM-Span approach can meet timeliness requirements and provide only the results of business interest.

Retention of Customers. The mobile Internet service market is very competitive, and providers should strive to retain customers and facilitate their continued usage of mobile Internet services. The expectancy confirmation model (ECM) was used to explain consumer behavior to continue using e-commerce services. A study with 244 survey respondents extended ECM to explain consumer intent to continue using mobile commerce services. Survey results showed that trust and satisfaction influenced the repurchasing activities of mobile shoppers. Trust involved the privacy and security of mobile commerce transactions, along with the quality of the purchased products or services. Satisfaction suggested how the expectation of goods or services performance were improved by mobile commerce transactions.

Mobile commerce providers should leverage the factors of perceived usefulness, trust, flow experience, and switching costs, which all significantly influenced the continued customer usage of services and ultimately determined their intent to change providers. Perceived usefulness reflected the advantages of ubiquity and immediacy with mobile services utilities. Trust was built upon the consumers' beliefs about the supplier's ability, integrity, and concern for the customer. The flow experience was the holistic sensation that people felt from the mobile Internet service. Their feelings were influenced by enjoyment, the concentrated immersion into the service, and control over the activity.

The cost of acquiring a new customer was five times more than the cost of retaining an existing one. The low switching costs enabled consumers to switch providers quickly if they were not satisfied with their current services. Mobile service providers needed to increase their current users' resistance to change to help facilitate their continued usage. Customers who resisted change more than likely retained relationships with their current service providers.

Case Analysis: Nokia

Nokia is a communications and information technology company providing Internet services. These services include applications, games, music, media, messaging services, and digital map and navigation services. Nokia also provides telecommunications network equipment and services. They had been the world's largest mobile phone manufacturer until 2012 when their old technology succumbed to the smart phone era. As of 2013, Nokia was comprised of 90,000 employees in 120 countries around the world. They generated revenues of an estimated $19.6 billion in more than 150 countries.

Mobile Phone Industry Transformation

Nokia, Samsung, LG, and Motorola were by volume among the largest mobile phone companies in the world during the feature phone era from 1994 to 2004. However, introduction of the smart phone in 2007 dramatically changed the competitive landscape of the mobile phone industry. Nokia was the first mover into the U.S. smart phone industry, but Apple followed to overtake this segment as the smart phone leader. Newcomers such as Google and Microsoft also posed a competitive threat to further shrink Nokia's prominence. This industry has become too competitive for Nokia with Apple, Lenovo, LG, and Samsung and newcomers such as Google and Microsoft.

From a global perspective, the new players included Apple and HTC. Samsung adopted smart phone strategies, but Nokia remained with their current profitable Symbian technology that eventually faded in market shares. The industry is currently dominated by companies such as Apple, LG, Nokia, and Samsung. Of these companies mentioned, only Nokia did not embrace the new smartphone technology. Nokia underestimated the threat of the smartphone and did not fully understand the competitive threat of this disruptive innovation to the mobile phone industry.

Digital Business Ecosystem

The evolving dynamics of the mobile phone industry from simply a feature mobile phone to more of a handheld computer known as the smart phone can be attributed to a digital business ecosystem. This ecosystem is influenced by how mobile phone providers, service providers, mobile users, individual mobile app developers, relevant companies, and digital services are connected and interact. The Nokia failed to transition their Symbian-based phones towards this ecosystem. However, their competition, such as the Apple and Android phones, embraced this ecosystem by providing better graphical user interfaces, better support, and a variety of phone software applications, known as apps. Consumers expected and demanded smartphones with attractive and user-friendly interfaces and Nokia did not meet their demands like the competition. Smart phone shares within the mobile phone market have grown from 5 percent in 2007 to 36 percent in 2014. Nokia is currently in a renewal stage with their dying Symbian ecosystem for their once profitable dumb phone. Their old Symbian operating system has led to a dramatic loss in market share. Nokia will need to analyze the current relationships within their ecosystem.

Change in Company Focus

Nokia was the biggest seller of Windows phones, but also the slowest competitor to adapt the new smart phone technology. Microsoft and Nokia have merged their resources and capabilities into joint alliances over the years following the inception of the smart phone. This led up to Nokia's mobile phone division being a logical next step for Microsoft to acquire in 2013 if they were to be a future contender within the mobile phone industry alongside leaders such as Samsung and Apple. In September 2013, Nokia began the process to sell to Microsoft what was formerly the world's largest vendor of mobile phones. This $7.17 billion deal was finalized in 2014.

Nokia's pricing strategy had been similar to Samsung. They offered a product range from low to higher price phone models. Nokia not only made some strategic errors with innovation, but in their operations, as well. Their mobile device factory in Jucu, Cluj County was projected to manufacture Nokia Classic devices for Africa and Asia. However, this facility eventually failed and operations have moved to Asia. Capital budgeting and finance decisions will need to be made by Nokia if they are to grow in the future.

Capital Budgeting Perspective

From 2012 to 2013, Nokia's gross margin for their continuing operations increased from 36.1 percent to 42.1 percent. A higher NSN gross margin attributed to this increase. This higher margin was attributed to better efficiencies in Global Services, product mix improvements with higher margins, and divestment of less profitable non-core businesses. Nokia has a superior delivery quality strategy that enables them to respond efficiently and in a timely manner to meet customers' expectations. Their supply chain management has contributed significantly to their growth. Nokia should remain on track with this strategy to sustain their gross margin.

Nokia currently has three businesses: Nokia Here, Nokia Networks, and Nokia Technologies. Nokia Here focuses on the research and development of location intelligence, location-based services, and local commerce. Nokia Here, previously Nokia Maps, is currently considered one of the top geographical localization services used by popular companies such as BMW or Garmin. Nokia Networks focuses on mobile broadband and global services. It underwent restructuring from the divestment of their optical network business and the joint venture between Nokia and Siemens. Nokia Networks has now become the company Nokia Solutions and Networks (NSN), which offers mobile and fixed network infrastructure solutions, communications and networks service platforms. Nokia Technologies rely on their intellectual property and licensing. It looks to produce the next-generation smart mobile devices with virtual reality. Nokia Technologies plans to use a low-cost graphene material for building disposable high-tech devices that could disrupt the industry. These capital budgeting decisions may enable Nokia to compete again in the mobile phone industry. However, financial challenges remain for Nokia.

Financial Perspective

Nokia shares have declined 5.9 percent from about a four-year high. NSN is Nokia's biggest business and primary reason for this decline. NSN's adjusted earnings of 9 cents per share in the fourth quarter of 2014 did not meet the average estimate of 9.5 cents. Their earnings are expected to continue declining for this year. In 2013, NSN's adjusted operating margin was 9.7 percent. This margin has now increased to 12.2 percent, accounting for an estimated 90 percent of sales. Nokia predicts the margin for this year to reach 8 to 11 percent.

Increased sales from Nokia's other two businesses, Nokia Here and Nokia Technologies, should help bolster revenue and profitability to improve Nokia's overall debt rating, resurrecting it out of junk status. The research and development division in Nokia Technologies received favorable reviews from the introduction of their own tablet computer in November 2014. The operating margin for this division has decreased from 67 to 52 percent. More money will be spent on licensing activities and growth opportunities. This extra spending could reduce Nokia's operating profit for this year by as much as 2 percent.

A new capital structure strategy may help defray this extra spending and retain more profit. More equity actually from within Nokia itself will be used towards the investment in their growth opportunities. Nokia's equity program for 2015 will entitle their own employees to contribute a part of their salary to purchase shares in the company. This program will support the participants' focus and alignment with Nokia's strategy and long-term sustainability. Nokia appears to be a financially stable company with a debt-to-equity (D/E) ratio of .32 as compared to their industry average of .39. Their continuance to use more equity than debt for financing will contribute to their future stability.

CHAPTER THIRTEEN:
GLOBAL OPPORTUNITY

You must be relevant to make some money. Successful companies with a competitive advantage tend to align their distinctive capabilities with complementary market opportunities. These capabilities are derived from the identification of their foremost competencies uniquely blended into an inimitable value available for consumers with matching demands. They will understand how to create value for their customers. Their value chain should be broad enough in scope to adapt for any changes, but narrow enough for strategic decisions to stay focused on the primary objectives. Continual improvements along the value chain will optimize efficiencies and save money, thus, remain relevant. Any improvements will involve sourcing decisions. Companies may decide to outsource, whether it is onshoring or offshoring. Or they may decide to insource, but build new facilities abroad. The geographic location of distribution and production facilities could potentially affect cost savings.

In the year 2003, a Goldman Sachs economist conceived the acronym BRIC, which represents the successful emerging economies of Brazil, Russia, India, and China. These countries are all considered as potentially fast-growing markets. These geographic locations may be ideal for targeting potential market sales or for hosting value chain facilities. However, these countries may differ in political climate, market size and the balance of their economies. China may show the greatest potential opportunity, but could encounter the most challenges.

The global economic landscape requires an organization to exploit new opportunities in other countries. They will need to address the challenges from their diverse cultures and markets. The more successful companies will exercise a global mindset when they interpret the dynamics of global operations. This mindset enables firms to better acknowledge and embrace the cultural and market differences of other countries into their operations.

The organization's global awareness will determine the extent they can adapt their thinking with other diverse thought processes. This collective mentality can be reshaped by encountering new experiences, changes in influential personalities and social processes, or an overall employee turnover with an influx of new mindsets.

The mindset of the organizational culture is determined by the common programming of the employees. They are collectively conditioned by shared career experiences, lessons, successes and failures. The culture of an organization is not effectively influenced by how the tangible assets are managed. Rather their team spirit is driven by how managers can influence the intangible symbols commonly meaningful to the majority of employees.

Case Analysis: Ikea

Ikea is an example of a successful international company with a global mindset. Their culture has embraced the challenges across borders in leading fashion. Ikea is the world's largest e retailer that specializes in the sale of self-assembly furniture. This multinational company spawned from the mind of a 17-year-old Scandinavian by the name of Ingvar Kamprad dating back in 1943. Ikea is an acronym for Ingvar Kamprad Elmtaryd Agunnaryd, the name of his family's farm.

What started out as a mail-order business selling fountain pens, encyclopedias, table runners, udder balm and reinforced socks eventually grew to furniture sales five years later. Their range of products has vastly expanded into a wide variety of furniture types and designs for living rooms, bedrooms, kitchens, children's rooms, and entertainment centers. They also offer the following services: restaurants, home delivery, assembly, installation, kitchen planning, office planning and home furnishing consultation. Ikea's annual revenues and operating profits have been on the rise. Their annual revenues for 2013 totaled over $37,280 million, an increase of 3.2 percent over 2012 revenues. Their operating profits for 2013 totaled over $5,246 million, an increase of 15.2 percent over 2012 profits.

Currently Ikea is comprised of 351 stores in 43 countries across Europe, North America, Asia, and Australia. Their headquarters is located in Leiden, Netherlands (Collins, 2011). Ikea's European market accounts for 69 percent of their total revenue, followed by North America with 16 percent, Asia and Australia with 8 percent and Russia with 7 percent. They operate 32 distribution centers, 11 customer distribution centers, and 28 trading service offices in 24 countries. Ikea has leveraged over 1,046 suppliers across 52 countries in 2013. Ikea has been successful at sustaining a global competitive advantage for nearly a decade. This can be attributed in part to their strategies in operations, social initiatives and branding.

Operational Strategy

Ikea designs, manufactures, transports, sells and assembles their products. This enables them to perform more efficiently, leading to lower prices. Ikea depends on customer involvement to help save money, as well. Customers will choose, collect, transport and assemble Ikea products.

Ikea's global mindset enabled them to adapt to other markets such as in China. They were able to identify their strategic challenges for entering Chinese markets. The prices that Ikea charged were considered low in Europe, but high in China. They were able to leverage local stores and build factories in China to help lower overall supply chain costs. Further cost savings were realized by performing local quality inspections that were closer to manufacturing locations in China. These efforts cut prices by more than 60 percent, which enabled Ikea to offer the lower, expected price to Chinese consumers.

Corporate Social Responsibility Strategy

The economic and social context of the global competitive environment has evolved over the years. Business firms must take into account more than their profits and productivity. Corporate Social Responsibility (CSR) initiatives now come into play to do what is for the good of society. CSR efforts will contribute to sustaining a competitive advantage.

The morality of a country can influence a company's CSR mindset. Their social capital could drive this morality. The ethical factors that contribute to such capital includes: fairer distribution, better government, social cooperation, and economic ideals instilled in citizens. Moral beliefs and ideals may differ among countries. Their moral space defines their cultural-specific ethical principles that attribute to possible differences. The key will be to determine common ethical beliefs known as hypernorms that thread through the majority of societies

Shared beliefs about human rights would be a start of a successful convergence of countries within the global competitive environment. Sweatshops that abuse human rights still exist in all countries, but perspectives may differ in various cultures and markets. The challenge will be to establish an international standard to help regulate work practices, while taking into account of the reality of the differing moral spaces across the borders.

Ikea has proven to be a leader of CSR around the world. They have adhered to hypernorms that lead economic success with minimal environmental impact. Ikea has adapted somewhat to the differing moral spaces around the world. Their efficient use of human, technical and financial resources has contributed to CSR efforts. They have achieved sustainability from their products manufactured from renewable, recyclable, and recycled materials.

Branding Strategy

Ikea's CSR reputation has significantly strengthened their branding. Their corporate brand identity has successfully transferred across the borders. Ikea's branding has been representative of the following core values: togetherness, enthusiastic, renewal mindset, cost awareness, acceptance of responsibility, humble attitude, willpower, leadership by example and innovative.

A group of corporate managers at Ikea's headquarters encouraged brand comprehending. This improved the understanding of their brand. Culture and their core values were important in their brand comprehension. The culture reflects how things are done and the interaction among their people. This reflects what Ikea is about and lends to their reputation. Ikea's vision aims to create a better everyday life for people, which includes their consumers and employees. They offer a product to improve the lifestyle of the consumer and they offer the opportunity for their employees to grow professionally.

Ikea's brand is thoroughly documented in their governance and manuals. An expected code of conduct lends to their reputable brand identity. They enforce this code of conduct internally and require it externally with their suppliers. Their ethical culture has sustained impressive brand longevity.

Challenges

Standardization of their products has enabled Ikea to maintain a low-cost strategy, leading to their overall global competitive advantage. Unfortunately, such standardization limits the number of targeted market segments. Ikea has been conflicted between their continual efforts to reduce costs and maintaining quality products. There has been a rise of damaged, defective or returned products. Their competitors could monopolize on Ikea's shortcomings to provide higher quality, more customized products. Another concern similar to dilemmas faced by other large corporations such as Walmart, Ikea has received negative publicity for their poor treatment of employees and questionable marketing practices. This could affect their CSR efforts if the overall public consensus believes or agrees with the negative publicity. All of these issues in general could eventually damage Ikea's branding and decrease the loyalty of their customers.

Other low-cost retail giants such as Walmart are expanding into such primary Ikea segments as the homeware specialist market. Ikea should counter such threats indirectly by expanding into other segments so as not to rely on its current segments that are under threat. There is opportunity for Ikea to expand their grocery business into the growing demand of grocery products. This growth has been a result of more people inclined to eat healthier food at home versus eating the unhealthy food in restaurants.

Case Analysis: PIMCO - Allianz SE

U.S.-based Pacific Investment Management Company, LLC (PIMCO) located in Newport Beach, CA was founded in 1971 and became the world's largest bond investor while competing in two financial services submarket industries: The Open-End Investment Funds Industry and the Portfolio Management Industry. However, in 2000 PIMCO was acquired by Allianz SE, a German-based multinational financial services institution. This acquisition quickly added core competencies and capabilities to Allianz versus developing the same competitive assets internally, which would have taken a much longer time to become operational and capable of generating revenue. Allianz is based on a multidivisional structure that decentralizes control to their autonomous subsidiaries such as PIMCO. Currently as one of the world's largest bond investors, PIMCO currently have their footprints in both the Open-End Investment Funds Industry and Portfolio Management Industry.

The PIMCO Total Return Fund currently had assets under management worth $247.9 billion and was the world's largest mutual fund company until Vanguard overtook it in October 2013. The fund's assets under their management exceeded $287.7 billion. PIMCO has its footprints in the Open-End Investment Funds Industry and the Portfolio Management Industry. Their manager, Bill Gross, is renowned as the bond king. In 2011, Gross led PIMCO to bounce back during hard times, and PIMCO sustained its position as top bond investor until Vanguard surpassed it.

PIMCO's Strategy

PIMCO's deliberate strategy is to maximize total returns while preserving capital and investing in a prudent manner. They look to employ its proven investment process to help their clients seize the best opportunities under various market conditions. PIMCO's innovative products and strategies employ the firm's proven investment process to help their clients capture the best opportunities in various market environments. Their macroeconomic forecasting, authoritative sector and security analysis and rigorous risk management address the challenges of a rapidly changing world.

PIMCO's Business Model

The revenue generation of PIMCO relies on a diverse portfolio with at least 65% of revenue deriving from fixed income instruments, primarily bonds. Other diversifications for the remaining 45% of their portfolio include stocks, real estate investment trusts, credit default swaps and mortgage-backed securities. The cost structure denotes a low level of capital intensity within the industries. However due to regulations requiring the adherence to distribution channels, a substantial amount of capital is invested into investor centers to accommodate this requirement. The profit margin is at an acceptable level as long as total revenues cover operational cost to include the expense of required distribution channels and the margin sustains steady growth for the company.

Basic Strategy Execution Factors

There is no standardized magic formula involved when executing Strategy. However, there are some basic common factors that contribute to its successful execution. The outcome of strategy execution has a higher likelihood for success if an organization possesses the following factors:

- Managers and employees who can execute the strategy effectively
- Capabilities needed to execute the strategy effectively
- Organizational structure to support the strategy
- Sufficient resources are available to help execute the strategy effectively
- Policies and procedures to help guide the execution of the strategy
- Best practices and business processes in place to continually improve strategy execution processes
- Information and operating systems in place to enable a more proficient execution of strategy

- Incentives are directly linked to achievement of strategic and financial targets
- Corporate culture that promotes effective strategy execution
- Ability to leverage the leadership skills to move forward with strategy execution

PIMCO's strategic objectives align with the above factors and provide a supportive environment for a successful strategy execution. Next, PIMCO's actual performance within each of its industries will be reviewed, and then a SWOT analysis will define any strengths, weaknesses, opportunities, and any oncoming threats that will provide valuable information to help in any decisions to be made for PIMCO's way forward.

PIMCO within the Open-End Investment Funds Industry

PIMCO has taken 3.7% of the market share within the Open-End Investment Funds Industry, in comparison to the top three competitors' market shares of 11%, 6.6% and 6.1%. During the economically challenging past five years, PIMCO has provided value proposition to worried clients by offering safe investments through fixed income instruments. This enabled their industry-specific assets to grow steadily through this crisis period to $563 billion in 2013, despite dark times when other companies were on the decline.

PIMCO within the Portfolio Management Industry

PIMCO has taken 2.9% of the market share within the Portfolio Management Industry, in comparison to the top four competitors' market shares of 3.7%, 3.5%, 3.5% and 3.0%. During the economically challenging past five years, PIMCO has provided value proposition to worried clients by offering safe investments through fixed income securities. This enabled their assets under management to grow steadily through this crisis period to $2 trillion in 2013, while their revenues grew to a now $9.2 billion.

Strategic Alliances and Partnerships

A strategy map will dissect the overall strategies from all alliance partners into layers of strategic perspectives or themes and align all of the strategic objectives within each perspective. Each partner's strategy can be clearly communicated and merged into the perspectives to form a working collaborative strategy that will clarify expectations for all. It is vital that a collaborative strategy has been agreed upon or it risks failure down the road for any of the involved parties, if not all of them. A balanced scorecard will establish set targets for each collaborative strategic objective that can be measured for performance results. Following are some examples of successful alliances with PIMCO in the past few years.

In 2010, PIMCO partnered with Schwab to offer a laddered bond to wealthier clients (Fund Action, 2010). This same year, PIMCO entered into an alliance with Aegon to get their feet wet in the United Kingdom segment by offering to retail investors the PIMCO Select Fund via the Aegon investment platform.

In 2011, PIMCO entered into a strategic alliance with Skandia, the United Kingdom's leading investment platform, to launch two funds: Global Bond Fund and UK Corporate Bond Fund. The Global Bond Fund offers global fixed income, while the UK Corporate Bond Fund offers diversification benefits yielding greater potential return on investment. This collaboration infused significant growth and distribution within the UK.

In 2012, PIMCO overtook the European market as the top fixed income specialist with sales of $34.8 billion. They also enjoyed success with third and fourth place rankings for the Global Investment Grade Credit Fund with sales of $5.9 billion and Diversified Income Fund with sales of $5.6 billion.

SWOT Analysis

A SWOT analysis revealed the strengths, weaknesses, opportunities and threats for PIMCO. The key strengths of PIMCO have been their innovative products and proven investment process to seize the best opportunities under various market conditions. Their macroeconomic forecasting, authoritative sector and security analysis and rigorous risk management address the challenges of a rapidly changing world. Their global expansion within the European sector has proven very successfully.

Their current weakness may actually be their fixed income products, particularly bonds, within the U.S. industry. This is supported by the recent catastrophic losses in 2013 from some of their fixed-income options in intermediate bonds. During record withdrawals of taxable bond funds in 2013, PIMCO's Total Return Fund suffered the worst, at the hands of investors who withdrew $28.1 billion out of this intermediate bond strategy. Another intermediate bond category, PIMCO Investment Grade Corporate Fund, was among the worst performers in the summer of 2013.

Their opportunities will prevail in continuing success within Europe, but blue ocean territories in developing countries such as China will provide potential growth, however, with accompanying risks. Value innovation for self-help services and products will fulfill the growing needs within the Portfolio Management Industry.

Potential threats implicate how the dynamics of risk investment are changing from what appeared to be safe in the past may not necessarily be the same in the future. It may become of less risk to invest in a specific corporate stock versus in the debt of a rich European country. Bill Gross predicts that investment returns will not be as high as in the past. Due to the changing risks and lower returns, he emphasizes the importance to think differently about portfolios and risks.

Case Analysis: FedEx Corporation

The United Parcel Service (UPS) and Federal Express (FedEx) leads the global courier and delivery services industry with nearly 65 percent of the total industry revenue and while the U.S. Postal Service is actually a contender, it is officially not listed as a competing company within this industry since its primary function falls under the National Postal Service.

The FedEx Corporation is a U.S.-based multinational entity centrally located in Memphis, TN, and reaches around the world in over 220 countries and territories. Their corporate mission strives to maximize profits for vested shareholders by providing value-added logistics, transportation and related services through their operating subsidiaries and by developing mutually rewarding relationships with their employees, partners and suppliers, while keeping in mind the best interest of shareholder profit margins.

Their corporate strategy relies on a single recognizable brand around the world, leverages independent networks to meet the unique needs of various consumers and effectively collaborates managerial efforts to maintain loyalty with employees, customers and investors. The FedEx Corporation's business units consist of FedEx Express, FedEx Ground, FedEx Freight and FedEx Services.

FedEx Express, the largest of the four subsidiaries within the overarching Federal Express Corporate family, directly competes within the global express delivery market against UPS, DHL, and TNT Express. Their competitive assets include a fleet of 660 aircraft and 52,400 motor vehicles and trailers with the capabilities to deliver 4 million packages per day across the U.S. and worldwide to over 220 countries. FedEx Express strives for long-term sustainability by focusing on their strengths and their best bet for maximum growth with a goal to improve annual profitability to $1.6 billion by the end of fiscal year 2016. This will be achieved by five initiatives: staff efficiency, air fleet modernization, international profit improvement, domestic transformation and domestic yield management. It is also a priority to maintain trust with the customers, stockholders and employees.

A SWOT analysis revealed the strengths, weaknesses, opportunities and threats for FedEx Express. The key strength is their efficient hub and spoke network consists of 58,400 drop-off locations, 660 aircrafts and 52,400 vehicles or trailers providing the ability to deliver 4 million packages per day across 220 countries and territories around the globe. Another key strength is their powerful branding that is collectively recognized around the world. The key weakness is the decline of funds for the employee retirement benefits due to negative market performance. This is considered a weakness because liquid assets must be used to offset the benefits fund deficiency, diverting available cash funds for unexpected events. The key threat will be the unpredictability of fuel prices in the future.

Business Model

The FedEx Corporation utilizes a total global strategy (TGS) model similar to other multinational corporations, but they embrace it in different way. Their model is developed for global strategy, however, the infrastructure does not really reflect a global approach. In addition, their strategy development relies more on the industry drivers than input from its internal resources, however, key elements for global success are their organizational structure and people, both casting FedEx, confusingly, in both negative and positive lights. Thus, organizational structure and people mute each other and become, more or less, nonfactors.

The FedEx Express subsidiary leverages an efficient delivery system that points all network locations back to the central super hub, a network type referred to as hub and spoke, strategically located in Memphis. This hub was strategically located where the weather is rarely bad enough to close the airport terminal and the airport willingly upgraded their facilities to meet the requirements of FedEx Express. Both the aircraft and vehicle fleets were integrated into the same delivery system to maximize efficiencies and coordinated processes.

The unpredictable cost of fuel and maintenance for the FedEx Express aircraft used for deliveries has spurred the strategic initiative for modernization of the aircraft fleet. FedEx Express recently received a Boeing 767 freighter that provides improved fuel consumption and offers more efficient maintenance. The overall initiative will total 19 Boeing 767 aircraft by fiscal year 2019. Their inventory of 365 hybrid-electric vehicles has saved over 300,000 gallons of fuel since 2004. In a continual effort towards fighting fuel costs, FedEx Express upgraded their vehicle fleet with an additional 43 all-electric vehicles.

Customer value proposition. FedEx Express offers customers a value proposition with a variety of time-certain delivery services ranging from one to three days and domestic pickup and delivery services in 220 countries or territories around the world. They also offer a general delivery option in countries such as China with medium to large companies that require no specific time for arrival.

Profit formula. FedEx Express generated over 61 percent of the annual sales revenue for the Federal Express Corporation in the fiscal year 2012, showing a 7.9 percent annual increase from the previous fiscal year. With expected high operating costs in this industry and the low profit margins per delivery, FedEx's business model must depend on the high frequency of deliveries to provide enough revenue for covering costs and maintaining a profit margin acceptable to the shareholder. If costs rise, it will be important for FedEx to be capable of passing along any increases to the consumer so that it will not decrease shareholder profits. FedEx has been recognized for their employee-friendly policies that have transcended their employees into a more dedicated workforce who has gone the extra mile to win over a customer's loyalty. Loyal customers will be more inclined to remain with FedEx and pay higher prices passed along from any additional costs versus those customers who would jump over to the competition if similar services are available at a cheaper price. The FedEx corporate strategy relies on managerial efforts, along with the importance of their branding, to maintain loyalty with employees and customers.

Environmental and Competitive Trends

Industry growth correlates with consumer spending because the more people buy, the more demand will be needed to deliver the goods. The industry suffered revenue declines due to the economic downturn since 2009, but has finally realized a growth of 2.5 percent growth this fiscal year. FedEx actually displayed a revenue growth of 4 percent in 2013.

The industry trend expects 2 percent revenue growth in fiscal year 2014, following 1.8 percent annually until 2018, a monetary value of $93.3 billion. The sectors of retail, wholesale, health and finance, along with changes in technology, unpredictable fuel prices and competitive substitutes will influence improvement within this industry. The trend of substituting documents with more online services has impacted the industry and it is expected to continue. However, the impact will not necessarily remain negative overall in the future since e-commerce will more than likely increase delivery demands due to growing online orders from consumers. Tougher environmental regulations are expected to influence industry throughout the next five years, forcing competitors to increase alternative-fuels resources and capabilities.

UPS dominates the U.S. market, but FedEx leads the international market. However, UPS is continuing their efforts to expand in Asia-Pacific, Latin America and Europe segments through acquisitions. UPS has teamed up with Daimler Chrysler and the U.S. Environmental Protection Agency in the development of a zero-emission fuel-cell van to defray any future environmental regulatory concerns.

UPS appears to be researching improvements to their supply chain management and logistics, while developing e-commerce solutions. Their actual revenue growth for 2013 was 1.7 percent versus FedEx with their growth of 4 percent in this same period. Even though they are not considered officially in the industry, the U.S. Postal Service still poses to be a large-scale competitive threat via the U.S. postal system, coupled with their own online capabilities.

Resources and Generic Strategy Options

The FedEx Express subsidiary generates more revenue for the FedEx Corporation than all of the other subsidiaries combined; therefore, the majority of any reinvestment funds should support this business unit with fleet upgrades, improved systems and acquisitions for new FedEx Express locations around the globe. Since financial resources are limited, management decisions must comprise not only reinvestment allocations, but also periodical dividend increases for their shareholders.

Michael E. Porter offers five possible generic competitive strategies: low cost product in a global market, low cost product in a local market, product differentiation in a global market, product differentiation in a local market and a best cost approach. Currently, FedEx appears to be using the best-cost provider strategy, offering a wide range of services at a moderate price that could still compete on either a global or local scale. FedEx's has devised a new low-cost strategy within local markets in countries such as China, where it provides deliveries for medium to large companies that require no specific time of arrival.

FedEx leverages its brand as a perceived differentiation. The organization should continue this strategy in pursuit of any new markets. FedEx Express appears to be competitive both at the current time and in the near future, choosing to sustain innovation incrementally by upgrading its fleets with new technology.

Since the earlier SWOT analysis revealed unpredictable fuel costs as the key potential threat for FedEx Express, the company's leadership should hire fuel managers to control the fluctuation of fuel prices with such tactical moves as fuel hedging, implementing fuel surcharges, and maintaining distribution contracts with suppliers. This approach will keep costs fixed for at least a more fixed and predictable period of time.

Review of China

During the past few decades, globalization has dramatically expanded towards the fragmentation of production across various countries. An evolution in the global landscape, which began with only the trade in final goods, has led to the involvement of all stages along the supply chain from manufacturing to the end product. This progression has cultivated the rapid growth of Chinese exports.

Financial Implications

Lower Costs. China became the most popular among the BRIC countries to attract outside manufacturers because of their low production costs. China has kept these costs at a competitive price. Production costs include labor, manufacturing, raw materials, energy, and transport.

However, overall labor costs in China have now been on the rise. This will impact manufacturers such as in the clothing industry who will rely heavily on cheaper labor. However, chemical companies will rely more on cheap energy costs versus labor costs. They will not be impacted in the same way. The various types of production costs will determine what types of companies will reap the most benefit with respect to current cost rates in China.

Developing markets in China pose a new set of requirements that stimulate new innovation, otherwise missed if companies are limited to the more mature markets. KFC ensures most of their businesses in China will primarily rely on suppliers located in China. This keeps their costs low and builds a stronger relationship with the Chinese government.

Added Value. China has contributed to the low value-added stages of the global supply chain, but not the higher value-added, more profitable activities. They have also contributed to the low-skilled labor for the global supply chain, but not the higher skill labor. Firms have primarily retained their strategic and higher value-added functions either internally or onshore outsourcing within their home country. Their skilled workers and intangible capital are available in country. However, a shift towards more global availability within both developed and developing countries such as China may be occurring.

Role of Technology

Technology choices will be made that include using current equipment, upgrading the current equipment, replacing current equipment, adding additional capacity or purchasing new equipment. High-level leadership must strategically decide the amount of capital investments to be dedicated towards technology. These strategic decisions determine what funds are committed in the present that is expected to return future profits for these technology investments. Their selection process for any new technology will eventually either contribute to overall revenue or cost in the organization.

In past years, China offered an ample supply of inexpensive labor that would lend to the higher profit margins from manufacturing technology products. However, overall labor costs have risen continuously since 2006 and have diminished the advantage of low labor costs in China. Flextronics is one of the largest contract manufacturers in the world that is based in China. Up until 2006 they were able to rely on low labor costs to mass produce consumer electronics at a more than acceptable profit margin. As labor costs began to rise in China, their profit margin began to shrink to unacceptable low levels.

Flextronics aggressively pursued research and development efforts to innovate their business model and automated their processes. They improved the accuracy of manufacturing activities and revamped their training for workers to operate new equipment efficiently and within the new work processes. Flextronics transformed their product line from the lower-valued consumer electronics to the higher-valued machines for use in the aerospace, robotics, automotive and medical industries. These new products now contribute to 72 percent of their output for consumers.

The Chinese government endorses Flextronics' new business model for other manufactures in country to follow. The government has invested heavily in these advanced industries that has increased their research and development. As a result, science and technology have grown in China. The export of high-tech products have surged from $150 billion to $600 billion during the past decade. China has become the largest exporter of these technology products; by 2022, they will produce over 30 percent of the world's electrical products.

Capital budgeting techniques are used to assess the time value of money, risks, and benefits associated with capital investment decisions. Analysis involves the payback period, net present value, and internal rate of return. The financial analysis of technology often overlooks certain factors such as total purchase cost, cost savings, revenue contribution and risks.

The total purchase cost of new technology can involve much more than the initial cost for the equipment. Other associated costs may involve any additional requirements to enable a company to integrate effectively and use the technology. Such necessities may include special fixtures, specialized tools, installation, technological adjustments, maintenance, and training. Operational processes will elaborate the most efficient use for the new technology to be integrated.

Step-by-step procedures will ensure the optimal way to operate, start, stop, load, unload, change over, upgrade, network, maintain, repair, clean up, speed up and slow down new technology. Chinese companies have an immense knowledge base for mass production activities and the capability to advance technology and innovation for the affiliated processes to produce quicker, cheaper and of acceptable quality.

New technology can save on costs through higher quality and more efficient operations. Efficient processes can save on the amount of labor, machines and material used. The use of fewer machines will lower the risks of machine repairs, leading to less downtime for production. Higher quality products will lower the amount of inspections, reworks and scraps. Safety risks can diminish significantly and save costs associated with lawful compliance and fines for any violations.

New technology can improve the process flow that accelerates the delivery of the final products or services to consumers. This new level of dependable delivery can attract more customers, leading to the generation of additional revenue. The flexibility of new technology will lend to the sustainable retention of their consumers with changing needs.

New technology will present the risk of uncertainty. Only estimates can be determined for the capability, lifecycle, operating cost, and financial rate of return for any new technologies. The actual outcome may be exponentially lower or higher than anticipated. Consequentially, this makes managerial decisions the all more important, but often most difficult to make, especially when involving collaborative ventures with Chinese companies.

The ability for managers to make the best choices about new technology is extremely important for the future of an organization. Schiavone (2011) noted the significance of a prompt and effective reaction to technological changes. Such adjustments can be either a full fledge transformation or the substitution of old products with new ones. Creative destruction lends to this process through the recombination of existing assets and resources that leads to new innovations. Current technology decisions will determine the future of how a company aligns their distinctive capabilities with complementary market opportunities.

Business Models

The most successful companies involved with China may need to leverage an innovative business model combining two separate models: source-centric and sales-centric. Companies that leverage source-centric models will utilize the lower cost production and supply resources from developing countries, but export the finished products outside of that source country. A sales-centric model is involved more with how the company will focus on the local marketing and sales of the finished products in the developing countries. In the innovative hybrid model, a company will need to determine the appropriate target markets for the sales of their products based on price.

The larger U.S. enterprises have successfully managed their supply chain in China. They tend to have a flexible supply chain system that can adjust with any market development and changes. U.S. enterprises may need to decentralize control and allow each local store or facility to manage their own staff, commodity, product selection, pricing, promotions, orders, and merchandising. This will allow for a quicker response to the local market and customer needs. Decentralization can enable firms to seize opportunities of local markets, but risks would have to be managed.

Walmart's business model for their global retail success has leveraged a productivity loop relying on lean operations and low prices. However, constraints in China forced them to change their business model. Their decentralization of operations has presented Walmart with challenges and risks. Questionable practices by local store managers may have been hidden from top leadership at the executive level.

Constraints

Business Strategy. Companies involved with developing countries may need to implement a constraint-driven strategy into their business model. They will perceive any constraints within a developing market as opportunities. Innovation of their processes or products will be needed to overcome such constraints. Companies will need to embrace the indifferences in China and learn from them. They should find new ways to provide value for Chinese consumers. Companies can leverage these new ideas, targeting consumers in other developing markets and even bring these innovations back to their home country facilities. Constraints are transformed into strengths that will lead to a company's global competitive advantage.

Political Climate. Most of the Fortune 500 companies have a presence in China. However, they struggle to realize the same profit margins as in other countries. The Communist Party of China controls the economic freedom for any companies doing business here, forcing these firms to adjust their strategies in alignment with Chinese government constraints. Economic liberalization is slowly evolving as their government is allowing more freedoms. China has been an open market that has spawned many new businesses and has offered more business opportunities for both local and outside companies to make some money versus any of the other BRIC countries. Such opportunities have allured many competitors. The competitive environment has become fierce and not for the faint hearted.

The Chinese government still owns the more important organizations affiliated with resources, energy, finance, communications, and media. The political climate has created barriers for any foreign companies to enter the Chinese markets easily compared to other global markets and entry continues to be difficult. Companies will need to optimize cost-saving processes to balance out the expectations of lower profit margins in China.

This may be very difficult for many foreign organizations to achieve, but the successful entry into Chinese markets will create a powerful global competitive advantage. The domestic companies in China are more accustomed to the political constraints and will more than likely maintain a competitive advantage over their foreign counterparts.

Walmart appears to be one of the larger companies that have been successful with production in China. However, they may have failed in local sales within these same domestic markets. By changing their business model, Walmart faced a smaller consumer market and forced to sell at higher prices to make up for the lower profit margins. Since many of the poorer market segments in China are not attracted to higher priced products, Walmart has been performing questionable practices in bulk sales to other retailers. However, such practices are perceived in China as ethical, therefore, acceptable with management at local stores.

Innovation

One of the innovations that U.S. enterprises can bring back home from China involves the improvement of speed and cost. The abundant supply of engineers has enabled Chinese companies to accelerate their research and development growth. They have industrialized innovation by applying what they have learned from their own extensive experiences with production lines. Lead times for new product development have been decreased significantly. For new design and development, faster user feedback has improved the learning curve for any new processes. Chinese organizations have been restructured to solve problems more rapidly.

Other companies such as those located in Silicon Valley may be following similar innovative paths. However, Chinese companies appear to be successfully achieving accelerated innovation that can produce quicker, cheaper and of good enough quality. This offers new capabilities that will lend to a global competitive advantage.

Strategic Alliances

Companies entering China have faced great difficulty in retraining their staff for adapting to a new environment. Employees will need to change their biases and vested interests. They are faced with language barriers and cultural differences. Sales and marketing personnel may not understand the religious or political taboos that exist. Even if they understand all of these elements in one geographic location of China, it could be completely different in another in-country location. Alliances and coalitions with Chinese firms could improve retraining efforts, better transcending the organizational understanding of Chinese culture.

Successful U.S. firms have established strategic partnerships that promote coexistence and co-prosperity with their key suppliers in China. Collaborative alliances with companies in China have enabled Walmart to leverage their innovative business model that is both source centric and sales centric. Future U.S. firms may not need to leverage the exact type of business model, but they will need to consider similar strategic alliances in creating value within the global supply chain.

U.S. firms have leveraged strategic alliances to maneuver more easily into Chinese markets. Service alliances enable U.S. firms to obtain tacit knowledge about local Chinese culture and markets that improve delivery of services. U.S. manufacturing alliances transfer knowledge to local Chinese partners in exchange for local market accessibility and lower operating costs. Big businesses such as KFC, L'Oréal, VW and Adidas have all hired the best local talent in China so they can better understand and cater to the Chinese customers.

Smaller companies will tend to partner with other organizations. Partnerships may introduce opportunities for small businesses to build upon their core competencies, speed up production to market or save on costs. China has global reputation for faster production at cheaper costs. All of these possible outcomes from partnerships in China can lead to innovation and ultimately, a global competitive advantage.

The potential downside for alliances involves the risks of leaking intellectual property outside the organization. Companies that rely on their unique flavor of knowledge could be severely damaged from such informational compromises. Knowledge management will be tested for these types of challenges.

Market Segments

The Chinese consumer market is comprised of many regions growing at different rates and on different paths. Successful companies will categorize the different local Chinese markets according to consumer preferences. Managers will cater to the segments with preferences that align with the capabilities of the company.

It will be important for companies to market not only to the coastal cities of Shanghai and Beijing, but in the lower-tier cities across the entire country. Samsung, L'Oréal and Adidas have been very successful with this type of marketing strategy. They have found the attitudes to vary in the different regions of the north and south. A growing number of consumers in the inland cities are attracted to the aspirational foreign brands.

In the past, the majority of Chinese consumers valued low prices over all other product factors. However, there has been a rise in the standard of living for more sectors in China. This has created a mix that introduces a change in flavor for the wealthier consumers who prefer the higher quality international brands, instead of just the cheapest. Higher end products will be sold to the wealthier sectors in China and exported to other wealthier countries. Sales for the poorer local markets in China will remain lower end products.

Companies are learning to adjust with the changing social dynamics in Chinese markets. According to top companies in China such as the McDonald's corporation, the successful businesses are in tune with the changing social attitudes of their customers in the local Chinese markets. They are aligning with the continuously rising aspirations to value and to be valued more. Firms will need to provide value that is relevant to the changing needs of Chinese consumers. Products that cater to the needs of other countries may not apply in China. While households in other countries may use three buckets to wash clothes, households in China use five buckets. Unilever, taking this difference of habits into consideration, has successfully promoted its Omo brand detergent in China.

China has focused to maintain healthier and more stable development of their local retail enterprises. The domestic retail enterprises in China will need to establish internal controls that can improve efficiency, reduce operational costs and increase profits. China has presented challenges for U.S. enterprises to control logistics distribution with a long-term advantage.

Social Implications

China has been known to be a haven for sweatshops, which are defined by the U.S. Department of Labor as factories that violate labor laws such as poor working conditions, unfair wages, unreasonable hours, child labor, and lack of benefits for workers. Companies from mature countries are often concerned with their corporate social responsibility (CSR) initiatives. Their moral obligation may go above and beyond what is mandated by law to do what is best for the people's welfare and the environment. However, this viewpoint may not align with the work practices in China. CSR relies on the premise that if an organization can enhance their self-image, brand awareness and reputation that they will appeal to more stakeholders. CSR can impact the customer's attitude towards a product, their intent to purchase and their loyalty to a product or organization.

Within the past decade, China has become more than a cheap source for production, often with a sweatshop reputation. Supply chain engagement has expanded beyond just production. Companies have increasingly catered to the local consumer markets, as well. During this period, Chinese citizens have increasingly become more concerned about CSR issues within their market segments. They continue to question multinational companies in China about environmental matters such as pollution or social problems about sweatshop conditions.

More than 80 nongovernmental organizations now exist for the purpose of environmental and social affairs in China. Companies that sell in Chinese markets should take CSR issues as serious here as if they were in the more mature countries such as the U.S. Manufacturing companies that will focus more on just the production side of business may not be pressured to focus on CSR concerns as much. However, CSR initiatives could transform business models that enable manufacturers to achieve and sustain a competitive advantage for the longer term.

Many of the smaller manufacturers in China have still not adopted CSR practices into their culture and work processes. For those that have attempted to integrate CSR into their business models, it has been with a short-term vision. This myopic approach has not really paved a true path for how CSR can lead to sustainability for them.

Small family-owned type of manufacturers are more concerned with how to make needed profits, many of which may not be very efficient with their operations. Owners may not understand how to pursue CSR initiatives or even what it really means. U.S. enterprises will need to proceed cautiously when dealing with manufacturers who have not adopted CSR practices. Times have changed and consumers are more social conscious. The sweatshop reputation will not bold well with many consumers in the more mature, developed countries.

CHAPTER FOURTEEN:
INTERDEPENDENT LEADERSHIP

You must be relevant to make some money. A dependent culture within an organization represents more of centralized decisions made by positions of authority. An independent culture represents more of decentralized decisions made by leadership positions, which emerge from the need of their competence. An interdependent culture represents a collective leadership (McCauley, et al, 2008).

The structure of an interdependent culture within an organization is treated as a whole instead of the recognition of the compartmentalized pieces that would otherwise strive to be distinct from each other. This interdependence promotes the perceptions of those who look at the organization as a whole instead of separate departments (McCauley, et al, 2008).

The Center for Creative Leadership described ten characteristics of Interdependent leadership discovered in their study (McCauley, et al, 2008):

- Organizational structure
- Executive level operations
- Organizational planning
- Lateral integration capability
- Performance-based pay and rewards
- Socially responsible initiatives
- Statements of organizational values
- Decisions making and problem solving
- Dealing with differences and conflict
- Facilitation of organizational change

An interdependent organizational structure will improve communications within the organization as a whole with no boundaries or hierarchical perimeters to constrict information and decision flows. Organizational resources and capabilities can grow and be used for optimal planning and implementation of the best solutions. Within the pay and benefits system, good ideas can be recognized and contributors rewarded fairly for their perspectives, knowledge and new ideas. It is important to know an organization's vision, beliefs and principles (McCauley, et al, 2008).

It can be even more important to exhibit social responsibility initiatives, to reflect a purpose beyond the scope of the organization in helping others within our society. This instills trust, loyalty and motivation throughout the organization. It should be expected for differences to exist among so many diverse personalities within an organization. Discussions should openly invite all viewpoints and address conflicts with encouragement for participation. Change is imminent and an organization should be adaptable to such changes for the sake of its future existence. Leaders will need to facilitate teams and processes to support any transformation with optimal results (McCauley, et al, 2008).

Importance of Collaboration

There could be an assumption made that the younger generations will more effectively discover new technologies or innovation that keeps pace with today's demands. These generations have grown up surrounded with the internet and other technologies. However, it can be a mistake to only rely on them to move forward with new technologies and innovative ideas. It will require a change in the organization's culture to involve a team effort throughout the organization. Collaboration is perhaps the most important single element that will determine how leaders and teams can effectively progress in the present, while impacting the future of an organization.

The underlying thread throughout the previously mentioned characteristics is collaboration. The interdependent culture relies upon the collaboration across an organization that has no authoritarian department boundaries or chain of command requirements. It fosters the cross-utilization of organizational resources and capabilities to discover innovative processes and results. The freedom to interface directly across departments with differing skillsets can provide a variety of expertise in a working group that increases the chances of finding optimal new solutions that may not otherwise be found by a single department with a more limited range of expertise.

Collaboration not only helps to tear down boundaries, but it can foster the creation of new space within an organization. This newborn place can incubate an exchange of freshly grown ideas that is vital for the successful outcome of collaboration. Leaders can help massage and stimulate new thoughts and create new knowledge within this new space with no boundaries. Solutions can be found and growth can be realized. An established collaborative culture can enable sustainment of innovation to promote a competitive advantage for now and beyond.

This collaborative culture can evolve into something even more prosperous by transcending from within an organization to outside of it in the form of partnerships or joint ventures with other organizations. This can significantly increase the resources and capabilities of all stakeholders in pursuance of something new that may have been unconceivable before the collaboration among organizations. However, organizations should cautiously align with other organizations of similar cultures, values and beliefs, otherwise the challenges may be far too great for a successful outcome of the collaboration.

Leadership Development Theory

According to the dynamic theory of leadership development, the ideal leadership and ethics does not exist within a person. Any pursuit to obtain these ideals are actually impossible to achieve, however, it can drive the forces of leadership development towards improvement.

Adaption-innovation theory explains how people's cognitive styles fall along a continuum between adaption and innovation. They usually lean more towards one or the other when solving problems. Adaptors will use known standard techniques to approach an issue and innovators will use fresh new ideas that could be out of the box. This theory also discusses how it is important to balance between both divergent and convergent thinking when groups are brainstorming new ideas. People tend to feel a need to be part of a group and maintain a group consensus on the best course of action; however, this can limit independent thinking that could actually foster additional resolutions. It may be important to take this theory into consideration when planning a leadership development for each person.

The constructive-developmental theory (CDT) focuses on the progression of how a person can better understand themselves and the world around them. Cognitive and social development will provide the resources to more effectively lead and manage on a path leading towards long-term sustainability. An independent leader will take this path by delegating, enforcing accountability with their people and influence their actions through rewards and recognition of their expertise. These type of leaders will look past a problem and examine the underbelly of what may be causing the problem. The interdependent leader will act as the organizational change agent who can easily adapt and persuade others to be more flexible and responsive. They will be comfortably capable of addressing any conflicts that may arise.

CONCLUSION

You must be relevant to make some money. No matter if you work for a business or you are a business owner. Some of the ways discussed in this book to increase your chances to be relevant included trends, competitive advantage, innovation, business models, ethics, social responsibility, risk management, mobile technologies, global opportunity, and interdependent leadership.

Trends

A trend is the general direction in which something is developing or changing, such as in the fashion industry with the hottest styles. However, by the time a trend is noticed, there could already have been a missed opportunity to gain a competitive advantage in the market. Trends start out with a novelty (something new, original, or new) that is less noticeable and harder to detect. Novelties more often become fads and less often become trends. People who have the ability to discern between the two before it happens will indeed be ahead of the game and increase their competitive advantage and relevance.

Mega trends are the underlying transformative currents driving trends and influencing the global expanse of activities and processes shaping society. The key is for businesses to be aware of mega trends and to recognize any potential convergences of unrelated trends that could make what currently appear to be irrelevant trends for businesses to become relevant for them in the future. Individual empowerment, diffusion of power, sustainability, and technological evolution are the identified mega trends that will reshape the global business landscape in the next 15 to 20 years. The most important of these trends is individual empowerment, which is infused throughout most of the other trends. Within the next few decades, the spirit of individual initiative will help to diminish poverty, build a global middle class, expand access to education and improve healthcare.

Competitive Advantage

Organizations such as Exxon Mobil strive to gain and maintain a competitive advantage within their respective consumer markets. Some industries may maintain a single leader, while in other industries more than one company may have found a competitive edge, enabling them to surface towards the top. Different companies within the same industry can actually achieve and sustain a competitive advantage without directly disrupting one another. The ideology of market segmentation explained how this phenomenon could occur.

A company can achieve a competitive advantage with skills or resources it asserts to offset its competition. Trained personnel with unique skillsets that produce a specialized product could catapult the company past its competition. Resources such as ownership of a distribution facility or bargaining power with outside distributors would lower costs within the distribution chain, enabling a company to gain the ability to offer a more economical product to consumers. Organizations should choose resources with comparative advantages to help achieve a competitive advantage. The sustainment of this competitive advantage will rely on distinct, immobile resources that cannot easily be obtained or copied by the competition.

Innovation

The majority of businesses tend to focus naturally on keeping up with their competition, without exploring innovative ways or ideas to invent something new for the consumer. This may be due to their limited resources or a decision driven by the organization's culture. Managers may be more comfortable with conventional strategies to build upon the existing market, instead of leaders looking for a new spice that would most certainly entice the hungry to want more of this delightful fresh flavor. In the business world, innovation is a process intended to achieve a competitive advantage over competitors. Innovation can improve upon an existing product, service or the organization, itself, with an incremental approach or create something new with a more radical approach.

Disruptive innovation, value innovation and blue ocean theories emphasized the importance of being first to find something new to offer consumers that is unique enough by what is offered and the process involved to offer it that will thwart off any threat of a copycat. However, the value innovation theory focused on the elements of the current market and determines what values could be changed, added or eliminated that will result in a new market value. The fast second approach was the only approach mentioned that resonated with the advantage of being a copycat if the follower responds rapidly, taking over and dominating the new market. It was argued and a matter of perception whether a market can truly be a blue ocean or more of a gradient of red to purplish conditions. It can also be left to different interpretations whether a new value is actually blue ocean, disruptive, value innovation or more of a vast incremental improvement of an existing market.

Business Models

While considering the faster pace of change and innovation within a competitive market or recognizing oncoming catastrophe before the competition, business model innovation has become more prominent as a way to remain relevant by increasing the agility and speed of organizations to change how they create value. New business models should maintain customer loyalty, while building barriers to market entry. The intent is for customers to continue buying more than once. Relationship building with the customer through various means such as automatic renewals or product/service/support bundles can help sustain the longevity of customer business. The idea is to find ways that make it harder for customers to abandon the organization for an alternative solution offered by a competitor.

A new business model should contain mechanisms that enable an organization to reexamine assumptions used for the model. Assumptions should be driven from data gathered from various filtered and unfiltered sources. Diverse perspectives from within the organization and from outside (i.e. competitors or customers) will exhaust the possible scenarios, leading to a more robust business model. Organizations should form a portfolio of business prospects they can experiment with the new model and determine the most likely candidates for future opportunity.

The following three macro-dimensions should be considered when building a business model: Value proposition, value network, and financial configuration. The first macro-dimension considers the following parameters: platform characteristics, offer positioning, platform provisioning, additional services, and resources and competencies. The second macro-dimension considers the following parameters: vertical integration, and customer ownership. The third macro-dimension considers the following parameters: revenue model, and cost model.

Ethics and Social Responsibility

Honest and moral companies will gain loyalty with their customers, thus remain relevant. An aggressive milieu can often tempt organizational leaders to ignore, condone or even encourage immoral behavior if forging a competitive edge over rival companies. They rationalize such conduct with their belief that the ends justify the means. The leader will need to avoid such temptations and sew a moral fabric into the inner lining of their organization. Moral leadership and their available mechanisms contribute to the ethical orientation of an organization, impacting their strategy formulation. Ethical choices by leadership must consider a competitive return for their shareholders, treat employees fairly, minimize any harm to the environment, and work in ways that do not damage the communities in which it operates. This is known as corporate social responsibility and societal admiration of such a company attribute will help that company maintain relevance.

Risk Management

The success of risk management will hinge on the responsive actions of the leadership. Inactions may result in managers who focus too much on the positivity and provide only the answers that stakeholders want to hear without regard to all of the assessed risks. In other instances, managers may prolong any actions of assessed risks until they actually occur. Depending on the mission of an organization, the consequential aftermath may lead to lost opportunity, financial ruin, or even irrelevance from what could have been avoidable circumstances. An organization's tolerance for the uncertainty, or their risk appetite, can be driven by strategic choices or the ethical climate. A more benevolent ethical atmosphere fosters a higher propensity to engage with riskier behavior.

Mobile Technologies

Mobile technology has been one of the most pervasive information technology trends in the past 20 years. From 2012 to 2017, experts project that global mobile data traffic will reach an estimated 11 exabytes per month. The advantages of mobile technology and innovations were also accompanied by the disadvantages and challenges both consumers and businesses faced. Debates have risen and perceptions developed from the introduction of new mobile functionality, along with the benefits and hindrances to their use. The prolific popularity of mobile services has driven the surge of mobile device usage during what society is witnessing as the mobile Internet service era. A Cisco Systems study revealed that the number of mobile devices will reach 10 billion by 2017.

Consumers have come to expect their smartphones to provide all of their online needs, including entertainment, shopping, networking, traveling, and work. The downside to all of this demand was the amount of data traffic and power required to operate the mobile devices. Mobile traffic was expected to grow 13-fold from 2012 to 2017. Consumers have adopted widespread use of telecommunication technologies, mobile devices, and wireless software application that lead to mobile commerce activities. Businesses must examine the patterns of consumer user behavior related to mobile commerce. Their examination is essential for broad applications such as planning physical shopping sites, maintaining e-commerce on mobile devices and managing online shopping websites.

Global Opportunity

The global economic landscape requires an organization to exploit new opportunities in other countries. They will need to address the challenges from their diverse cultures and markets. The more successful companies will exercise a global mindset when they interpret the dynamics of global operations. This mindset enables firms to better acknowledge and embrace the cultural and market differences of other countries into their operations. The organization's global awareness will determine the extent they can adapt their thinking with other diverse thought processes. This collective mentality can be reshaped by encountering new experiences, changes in influential personalities and social processes, or an overall employee turnover with an influx of new mindsets.

Interdependent Leadership

The structure of an interdependent organizational culture is treated as a whole instead of the recognition of the compartmentalized pieces that would otherwise strive to be distinct from each other. This interdependence promotes the perceptions of those who look at the organization as a whole instead of separate departments. An interdependent organizational structure will improve communications within the organization as a whole with no boundaries or hierarchical perimeters to constrict information and decision flows. Organizational resources and capabilities can grow and be used for optimal planning and implementation of the best solutions. Within the pay and benefits system, good ideas can be recognized and contributors rewarded fairly for their perspectives, knowledge and new ideas. It is important to know an organization's vision, beliefs and principles.

REFERENCES

Anthony, S. D., Duncan, D. S., & Siren, P. A. (2014). Build an Innovation Engine in 90 Days. (cover story). *Harvard Business Review, 92*(12), 59-68. Retrieved from https://hbr.org/2014/11/build-an-innovation-engine-in-90-days

Auvinen, T. P., Lämsä, A., Sintonen, T., & Takala, T. (2013). Leadership manipulation and ethics in storytelling. *Journal of Business Ethics, 116*(2), 415-431. doi:http://www.dx.doi.org/10.1007/s10551-012-1454-8

Beeson, J. (2014). Five Questions Every Leader Should Ask About Organizational Design. *Harvard Business Review*. Retrieved from https://hbr.org/2014/01/five-questions-every-leader-should-ask-about-organizational-design

Bradshaw, K. (2013). Sourcing effective scenarios for use in business ethics training. *Industrial and Commercial Training, 45*(5), 264-268. doi:http://www.dx.doi.org/10.1108/ICT-01-2013-0002

Brondoni, S. M. (2015). Global networks, outside-in capabilities and smart innovation. *Symphonya*, (1), 6-21. Retrieved from http://symphonya.unimib.it/article/view/2015.1.02 brondoni

Christensen, C.M. (1997). The Innovator's Dilemma. Location: *Harvard Business School Press*

Christensen, C.M., & Overdorf, M. (2000). Meeting the Challenge of Disruptive Change. *Harvard Business Review, 78* (2), 65–76. Retrieved from https://hbr.org/2000/03/meeting-the-challenge-of-disruptive-change

de Brito, R., & Brito, L. (2014). Dynamics of Competition and Survival. *BAR - Brazilian Administration Review, 11*(1), 64-85.

Dustin, G., Bharat, M., & Jitendra, M. (2014). Competitive advantage and motivating innovation. *Advances in Management, 7*(1), 1-7.

Fourné, S. L., Jansen, J. P., & Mom, T. M. (2014). Strategic agility in MNEs: Managing tensions to capture opportunities across emerging and established markets. *California Management Review, 56*(3), 13-38. doi:http://www.dx.doi.org/10.1525/cmr.2014.56.3.13

Gao, T., Tian, Y., & Yu, Q. (2014). Impact of Manufacturing Dynamic Capabilities on Enterprise Performance-the Nonlinear Moderating Effect of Environmental Dynamism. *Journal of Applied Sciences, 14*(18), 2067-2072. doi:10.3923/jas.2014.2067.2072

García-sánchez, I., Rodríguez-domínguez, L., & Gallego-Álvarez, I. (2013). CEO qualities and codes of ethics. *European Journal of Law and Economics, 35*(2), 295-312. doi:http://www.dx.doi.org/10.1007/s10657-011-9248-5

Gupta, R. K. (2013). Core competencies for business excellence. *Advances in Management, 6*(10), 11-15. Retrieved from http://connection.ebscohost.com/c/articles/90615820/cor e-competencies-business-excellence

Hamdaoui, B., Alshammari, T. & Guizani, M. (2013). Exploiting 4G mobile user cooperation for energy conservation: challenges and opportunities. *Wireless Communications, IEEE, 20*(5), 62-67.

Hannachi, Y. (2015). Development and validation of a measure for product innovation performance: The PIP scale.

Journal of Business Studies Quarterly, 6(3), 23-35. Retrieved from http://jbsq.org/wp-content/uploads/2015/ 03/March_2015_3.pdf

Kim, C. & Mauborgne, R. (2004). Value Innovation: The Strategic Logic of High Growth. *Harvard Business Review, 82*(7/8), 172–180. Retrieved from https://hbr.org/2004/07/value-innovation-the-strategic- logic-of-high-growth

Magid, J., Sheskin, M. & Schulz, L. (2015). Imagination and the generation of new ideas. *Cognitive Development*, 34, 99 – 110. doi:http://www.dx.doi.org/10.10.16/j.cogdev. 2014.12.008

Marien, M. (2013). The provocative gt-2030 report: outline and comments. *World Future Review, 5*(4), 324-331. doi:http://www.dx.doi.org/10.1177/1946756713505360

McCauley, C.D., Palus, C.J., Drath, W.H., Hughes, R.L., McGuire, J.B., O'Connor, P.M.G., & Van Velsor, E. (2008). Interdependent leadership in organizations: evidence from six case studies. *A Center for Creative Leadership Report.* CCL No. 190. Retrieved January 27, 2014 from http://www.ccl.org/leadership/pdf/research/ interdependentLeadership.pdf

Mojica, I.J., Adams, B., Nagappan, M., Dienst, S., Berger, T., & Hassan, A.E. (2014). A large-scale empirical study on software reuse in mobile apps. *Software. IEEE , 31*(2), 78- 86. doi:http://www.dx.doi.org/10.1109/MS.2013.142

Musetescu, A. (2013). How to achieve a competitive advantage. *Knowledge Horizons.Economics*, 5, 13-16.

Parmar, R., Mackenzie, I., Cohn, D., & Gann, D. (2014). The New Patterns of Innovation. *Harvard Business Review, 92*

(1/2), 86-95. Retrieved from https://hbr.org/2014/01/the-new-patterns-of-innovation

Pisano, G. P. (2015). You need an innovation strategy. *Harvard Business Review, 93*(6), 44-54. Retrieved from https://hbr.org/2015/06/you-need-an-innovation-strategy

Raynor, M.E. (2011). Disruption Theory as a Predictor of Innovation Success/Failure. *Strategy & Leadership, 39*(4), 27–30. doi:http://www.dx.doi.org/abs/10.1108/10878571111147378

Salerno, M. S., Gomes, L. A. V., da Silva, D. O., Bagno, R. B., & Uchôa Freitas, C. L. T. (2015). Innovation processes: Which process for which project?, *Technovation*, 35, 59-70, doi:http://www.dx.doi.org/10.1016/j.technovation.2014.07.012.

San-Martin, S., & López-Catalán, B. (2013). How can a mobile vendor get satisfied customers? *Industrial Management and Data Systems, 113*(2), 156-170. doi:http://www.dx.doi.org/10.1108/02635571311303514

Schwarz, J.O., Kroehl, R., & von der Gracht, H.A. (2014). Novels and novelty in trend research. *Technological Forecasting and Social Change, 84*(5), 5–65.

Sekerka, L. E., Comer, D. R., & Godwin, L. N. (2014). Positive organizational ethics: Cultivating and sustaining moral performance. *Journal of Business Ethics, 119*(4), 435-444. doi:http://dx.doi.org/10.1007/s10551-013-1911-z

Shuen, A., Feiler, P., & Teece, D. (2014). Dynamic capabilities in the upstream oil and gas sector: Managing next generation competition. *Energy Strategy Reviews,* 8. doi:http://www.dx.doi.org/10.1016/j.esr.2014.05.002

Spaid, B. I., & Flint, D. J. (2014). The meaning of shopping experiences augmented by mobile internet devices. *Journal of Marketing Theory and Practice, 22*(1), 73-89.

Teece, D. J., Pisano, G., & Shuen, A. (1997). Dynamic capabilities and strategic management. *Strategic Management Journal, 18* (7), 509-533.

Tideman, S. G., Arts, M. C., & Zandee, D. P. (2013). Sustainable leadership: Towards a workable definition. *The Journal of Corporate Citizenship,* (49), 17-33.

Upadhyay, A. Y. A., Upadhyay, A. K., & Palo, S. (2013). Strategy implementation using balanced scorecard: Achieving success through personal values of leaders and employees. *Management and Labour Studies, 38*(4), 447-469. doi: 10.1177/0258042X13516596

Weigel, T., & Goffin, K. (2015). Creating innovation capabilities. *Research Technology Management, 58*(4), 28-35. Retrieved from http://psm.hongik.ac.kr/ezsboard_data/board/table_101/2_1430474794.pdf

Wessel, M. & Christensen, C.M. (2012). Surviving Disruption. *Harvard Business Review, 90*(12), 56–64. Retrieved from https://hbr.org/2012/12/surviving-disruption

Wu, Q., He, Q., & Duan, Y. (2013). Explicating dynamic capabilities for corporate sustainability. *EuroMed Journal of Business, 8*(3), 255-272. doi:http://dx.doi.org/10.1108/EMJB-05-2013-0025

Zacher, H., Pearce, L. K., Rooney, D., & Mckenna, B. (2014). Leaders' personal wisdom and leader-member exchange quality. *Journal of Business Ethics, 121*(2), 171-187. doi:http://www.dx.doi.org/10.1007/s10551-013-1692-4

ABOUT THE AUTHOR

Phillip E. "Dr. Phil" Copeland is a scholarly researcher and author within the field of strategy and innovation. He holds a Doctor of Business Administration degree in Strategy and Innovation from Capella University.

Dr. Phil has over 30 years of combined federal government and military service career experience in visual information management, project and program management, digital strategy, website design and management, information technology, training, and strategic communications. He is a retired Air Force enlisted leader who has always perceived problems as opportunities and found innovative solutions, often with limited time and resources. His positive, tenacious attitude has been the key to unlock the door between impossibility and possibility.

Dr. Phil is a member of both the Delta Mu Delta International Honor Society in Business and the International Association of Innovation Professionals (IAOIP). His dissertation research study investigated the adoption of mobile commerce by small businesses. This cutting-edge research pioneered new territory from a business perspective towards mobile technology, one of the most pervasive trends in the last few decades. The imprint of his research lays the path for others within both the academia and business worlds to follow.

www.ingramcontent.com/pod-product-compliance
Lightning Source LLC
Chambersburg PA
CBHW031934190326
41519CB00007B/523